T0016211

Programming Arduino®

Getting Started with Sketches

Programming Arduino®

Getting Started with Sketches

THIRD EDITION

Simon Monk

New York   Chicago   San Francisco
Athens   London   Madrid
Mexico City   Milan   New Delhi
Singapore   Sydney   Toronto

Library of Congress Control Number: 2022036835

McGraw Hill books are available at special quantity discounts to use as premiums and sales promotions or for use in corporate training programs. To contact a representative, please visit the Contact Us page at www.mhprofessional.com.

Programming Arduino®: Getting Started with Sketches, Third Edition

Copyright © 2023, 2016, 2012 by McGraw Hill. All rights reserved. Printed in the United States of America. Except as permitted under the United States Copyright Act of 1976, no part of this publication may be reproduced or distributed in any form or by any means, or stored in a database or retrieval system, without the prior written permission of the publisher, with the exception that the program listings may be entered, stored, and executed in a computer system, but they may not be reproduced for publication.

McGraw Hill, the McGraw Hill logo, TAB, and related trade dress are trademarks or registered trademarks of McGraw Hill and/or its affiliates in the United States and other countries and may not be used without written permission. All other trademarks are the property of their respective owners. McGraw Hill is not associated with any product or vendor mentioned in this book.

Arduino is a trademark of Arduino LLC.

1 2 3 4 5 6 7 8 9   LCR   27 26 25 24 23 22

ISBN   978-1-264-67698-9
MHID   1-264-67698-0

Sponsoring Editor Lara Zoble	**Project Managers** Poonam Bisht and Monika Chaudhari MPS Limited	**Indexer** Simon Monk
Editorial Supervisor Patty Mon		**Art Director, Cover** Jeff Weeks
Production Supervisor Lynn M. Messina	**Copy Editor** Subashree Baskaran MPS Limited	**Illustration** MPS Limited
Acquisitions Coordinator Elizabeth M. Houde	**Proofreader** MPS Limited	**Composition** MPS Limited

Information has been obtained by McGraw Hill from sources believed to be reliable. However, because of the possibility of human or mechanical error by our sources, McGraw Hill, or others, McGraw Hill does not guarantee the accuracy, adequacy, or completeness of any information and is not responsible for any errors or omissions or the results obtained from the use of such information.

To my boys, Stephen and Matthew,
from a very proud Dad.

About the Author

Simon Monk has a bachelor's degree in cybernetics and computer science and a doctorate in software engineering. He has been an active electronics hobbyist since his school days and is an occasional author in hobby electronics magazines. Dr. Monk is also author of some 20 books on Maker and electronics topics, especially Arduino and Raspberry Pi. Simon also designs products for MonkMakes Ltd. https://monkmakes.com

You can find out more about his books at http://simonmonk.org. You can also follow him on Twitter, where he is @simonmonk2.

CONTENTS

PREFACE

The first edition of this book was published in November 2011 and has been Amazon's highest ranking book on Arduino.

The Arduino Uno is still considered to be the standard Arduino board. However, many other boards, including both official Arduino boards (like the Leonardo, Nano, and Pro Mini) and other Arduino-compatible devices like the Raspberry Pi Pico, ESP32-based boards, and numerous Feather boards from Adafruit have also appeared.

The Arduino software is available for so many families of microcontroller, that it has become the environment of choice for many embedded programmers.

This edition also addresses the use of Arduino in Internet of Things (IoT) projects and the use of various types of display including OLED and LCD.

Simon Monk

ACKNOWLEDGMENTS

I thank Linda for giving me the time, space, and support to write this book and for putting up with the various messes my projects create around the house.

Finally, I would like to thank Lara Zoble and everyone involved in the production of this book. It's a pleasure to work with such a great team.

INTRODUCTION

Arduino interface boards provide a low-cost, easy-to-use technology to create microcontroller-based projects. With a little electronics, you can make your Arduino do all sorts of things, from controlling lights in an art installation to managing the power on a solar energy system.

There are many project-based books that show you how to connect things to your Arduino, including *30 Arduino Projects for the Evil Genius* by this author. However, the focus of this book is on programming the Arduino and Arduino-compatible boards using the Arduino IDE.

This book will explain how to make programming the Arduino simple and enjoyable, avoiding the difficulties of uncooperative code that so often afflict a project. You will be taken through the process of programming the Arduino step by step, starting with the basics of the C programming language that Arduinos use.

So, What Is Arduino?

The word "Arduino" has come to mean both the hardware and a software environment for programming microcontroller boards. Because microcontroller boards come in all shapes and sizes, our standard board will be the most popular official Arduino board, the Arduino Uno.

The Arduino Uno is a small microcontroller board with a universal serial bus (USB) plug to connect to your computer and a number of connection sockets that can be wired to external electronics such as motors, relays, light sensors, laser diodes, loudspeakers, microphones, and more. They can be powered either through a USB connection from the computer, a battery, or from a power supply. They can be controlled from the computer or programmed by the computer and then disconnected and allowed to work independently.

The board design of official Arduino boards and many Arduino-compatible boards is open source. This means that anyone is allowed to make Arduino-compatible boards. This competition has led to low costs for the boards and all sorts of variations on the "standard" boards.

The basic boards are supplemented by accessory shield boards that can be plugged on top of the Arduino board.

The software for programming your Arduino is easy to use and also freely available for Windows, Mac, and Linux computers. There is also a browser-based version of the software.

What Will I Need?

This is a book intended for beginners, but it is also intended to be useful to those who have used Arduino for a while and want to learn more about programming the Arduino or gain a better understanding of the fundamentals. As such, this book concentrates on the use of the Arduino Uno board, apart from Chapter 10 that uses an ESP32 Arduino-compatible board; however, almost all of the code will work unmodified on all the Arduino models and various Arduino-compatible microcontroller boards.

You do not need to have any programming experience or a technical background, and the book's exercises do not require any soldering. All you need is the desire to make something.

If you want to make the most of the book and try out some of the experiments, then it is useful to have the following on hand:

- A few lengths of solid core wire
- A cheap digital multimeter

Both are readily available for a few dollars from a hobby electronics store or online retailer such as Adafruit or Sparkfun. You will of course also need an Arduino, ideally an Arduino Uno and for Chapter 10, a low-cost ESP32 Arduino-compatible such as the Lolin32 Lite.

If you want to go a step further and experiment with displays, then you will need to buy those too. See Chapters 9 and 10 for details.

Using This Book

This book is structured to get you started in a really simple way and gradually build on what you have learned. You may, however, find yourself skipping or skimming some of the early chapters as you find the right level to enter the book.

The book is organized into the following chapters:

- **Chapter 1: Getting Started**   Here you conduct your first experiments with your Arduino board: installing the software, powering it up, and uploading your first sketch.

- **Chapter 2: C Language Basics**   This chapter covers the basics of the C language; for complete programming beginners, the chapter also serves as an introduction to programming in general.

- **Chapter 3: Functions**   This chapter explains the key concept of using and writing functions in Arduino sketches. These sketches are demonstrated throughout with runnable code examples.

- **Chapter 4: Arrays and Strings**   Here you learn how to make and use data structures that are more advanced than simple integer variables. A Morse code example project is slowly developed to illustrate the concepts being explained.

- **Chapter 5: Input and Output**   You learn how to use the digital and analog inputs and outputs on the Arduino in your programs. A multimeter will be useful to show you what is happening on the Arduino's input/output connections.

- **Chapter 6: Boards**   In this chapter we will look at the wide range of Arduino and Arduino-compatible boards to help you choose the right board for your project.

- **Chapter 7: Advanced Arduino**   This chapter explains how to make use of the Arduino functions that come in the Arduino's standard library and some other more advanced features of Arduino programming.

- **Chapter 8: Data Storage**   Here you learn how to write sketches that can save data in electrically erasable programmable read-only memory (EEPROM) and make use of the Arduino's built-in flash memory.

- **Chapter 9: Displays**   In this chapter, you learn how to interface an Arduino with displays and to make a simple USB message board.

- **Chapter 10: Arduino Internet of Things Programming**   You learn how to make the Arduino behave like a web server and communicate with the Internet using services.

Resources

This book is supported by an accompanying web page.

www.arduinobook.com

There you will find all the source code used in this book as well as other resources, such as errata.

1

Getting Started

Arduino is a microcontroller platform that has captured the imagination of electronics enthusiasts. Its ease of use and open source nature make it a great choice for anyone wanting to build electronic projects.

Ultimately, it allows you to connect electronics through its pins so that it can control things—for instance, turn lights or motors on and off or sense things such as light and temperature. This is why Arduino is sometimes given the description *physical computing*. Because Arduinos can be connected to your computer by a universal serial bus (USB) lead, this also means that you can use the Arduino as an interface board to control those same electronics from your computer.

This chapter is an introduction to the Arduino system including the history and background of the Arduino, as well as an overview of the Arduino Uno and Lolin32 Lite, the two Arduino boards that we will use in this book.

Microcontrollers

The heart of your Arduino is a microcontroller. Pretty much everything else on the board is concerned with providing the board with power and allowing it to communicate with your desktop computer.

A microcontroller really is a little computer on a chip. It has everything and more than the first home computers had. It has a processor, a small amount of random access memory (RAM) for holding data, some erasable programmable read-only memory (EPROM) or flash memory for holding your programs and it has input and output pins. These input/output (I/O) pins link the microcontroller to the rest of your electronics.

Inputs can read both digital (is the switch on or off?) and analog (what is the voltage at a pin?). This opens up the opportunity of connecting many different types of sensor for light, temperature, sound, and more.

Outputs can also be analog or digital. So, you can set a pin to be on or off (0 volts or 5 volts) and this can turn light-emitting diodes (LEDs) on and off directly, or you can use the output to control higher power devices such as motors. They can also provide an analog output. That is, you can control the power output of a pin, allowing you to control the speed of a motor or the brightness of a light, rather than simply turning it on or off.

The microcontroller on an Arduino Uno board is the 28-pin chip fitted into a socket at the center of the board. This single chip contains the memory, processor, and all the electronics for the input/output pins. It is manufactured by the company Microchip, which is one of the major microcontroller manufacturers. Each of the microcontroller manufacturers produces dozens of different microcontrollers grouped into different families. The microcontrollers are not all created for the benefit of electronics hobbyists like us. We are a small part of this vast market. These devices are really intended for embedding into consumer products, including cars, washing machines, TVs, cars, children's toys, and even air fresheners.

The Arduino system provides a standardized way of programming all manner of microcontrollers and is not limited to official Arduino boards. This means that whatever microcontroller you want to use, you can (with a few exceptions) program it as an Arduino without having to learn some manufacturer's proprietary software tool.

Development Boards

We have established that the microcontroller is really just a chip. A microcontroller will not just work on its own without some supporting electronics to provide it with a regulated and accurate supply of electricity (microcontrollers are fussy about this) as well as a means of communicating with the computer that is going to program the microcontroller.

This is where development boards come in. An Arduino Uno board is really a microcontroller development board that happens to be an independent open source hardware design. This means that the design files for the printed circuit board (PCB) and the schematic diagrams are all publicly available, and everyone is free to use the designs to make and sell his or her own Arduino boards.

All the microcontroller manufacturers—including Microchip, which makes the ATmega328 microcontroller used in an Arduino board—also provide their own development boards and programming software. Although they are usually fairly inexpensive, these tend to be aimed at professional electronics engineers rather than hobbyists. This means that such boards and software are arguably harder to use and require a greater learning investment before you can get anything useful out of them.

A Tour of an Arduino Uno Board

Figure 1-1 shows an Arduino Uno board. Let's take a quick tour of the various components on the board.

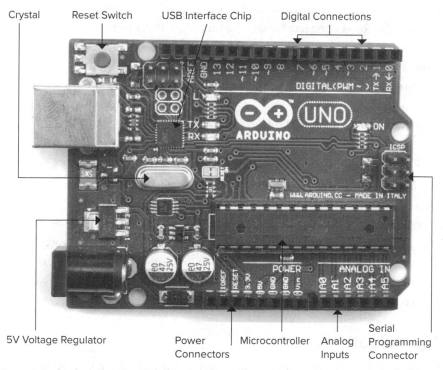

Figure 1-1 *An Arduino Uno board.*

Power Supply

Referring to Figure 1-1, directly below the USB connector is the 5-volt (5V) voltage regulator. This regulates whatever voltage (between 7V and 12V) is supplied from the DC power socket into a constant 5V.

The 5V voltage regulator chip is actually quite big for a surface mount component. This is so that it can dissipate the heat required to regulate the voltage at a reasonably high current. This is useful when driving external electronics.

Although powering the Arduino through the DC power socket is useful when running the Arduino from batteries or a DC power jack, the Arduino Uno can also be powered through the USB port, which is also used to program the Arduino.

Power Connections

Next let us look at the connectors at the bottom of Figure 1-1. You can read the connection names next to the connectors. The connector of interest is Reset. This does the same thing as the Reset button on the Arduino. Rather like rebooting a PC, using the Reset connector resets the microcontroller so that it begins its program from the start. To reset the microcontroller with the Reset connector, you momentarily set this pin low (connecting it to 0V).

The rest of the pins in this section just provide different voltages (3.3V, 5V, GND, and Vin), as they are labeled. GND, or ground, just means zero volts. It is the reference voltage to which all other voltages on the board are relative.

Analog Inputs

The six pins labeled as Analog In A0 to A5 can be used to measure the voltage connected to them so that the value can be used in a sketch (Arduino Program). Note that they measure a voltage and not a current. Only a tiny current will ever flow into them and down to ground because they have a very large internal resistance. That is, the pin having a large internal resistance only allows a tiny current to flow into the pin.

Although these inputs are labeled as analog, and are analog inputs by default, these connections can also be used as digital inputs or outputs.

Digital Connections

We now switch to the top connector and start on the right-hand side in Figure 1-1. Here we find pins labeled Digital 0 to 13. These can be used as either inputs or outputs. When used as outputs, they behave rather like the power supply voltages discussed earlier in this section, except that these are all 5V and can be turned on or off from your sketch. So, if you turn them on from your sketch they will be at 5V, and if you turn them off they will be at 0V. As with the power supply connectors, you must be careful not to exceed their maximum current capabilities. The first two of these connections (0 and 1) are also labeled RX and TX, for receive and transmit. These connections are reserved for use in communication and are indirectly the receive and transmit connections for your USB link to your computer.

These digital connections can supply 40 mA (milliamps) at 5V. That is more than enough to light a standard LED, but not enough to drive an electric motor directly.

Microcontroller

Continuing our tour of the Arduino Uno board, the microcontroller chip itself is the black rectangular device with 28 pins. This is fitted into a dual in-line (DIL) socket so that it can be easily replaced. The 28-pin microcontroller chip used on the Arduino Uno board is the ATmega328.

The heart—or, perhaps more appropriately, the brain—of the device is the central processing unit (CPU). It controls everything that goes on within the device. It fetches program instructions stored in the flash memory and executes them. This might involve fetching data from working memory (RAM), changing it, and then putting it back. Or, it may mean changing one of the digital outputs from 0V to 5V.

The electrically erasable programmable read only memory (EEPROM) is a little like the flash memory in that it is non-volatile. That is, you can turn the device off and on and it will not have forgotten what is in the EEPROM. Whereas the flash memory is intended for storing program instructions (from sketches), the EEPROM is used to store data that you do not want to lose in the event of a reset or the power being turned off.

Other Components

Above the microcontroller is a small, silver, rectangular component. This is a quartz crystal oscillator. It ticks 16 million times a second, and on each of those ticks, the microcontroller can perform one operation—addition, subtraction, or another mathematical operation.

In the top-left corner is the Reset switch. Clicking on this switch sends a logic pulse to the Reset pin of the microcontroller, causing the microcontroller to start its program afresh and clear its memory. Note that any program stored on the device will be retained, because this is kept in non-volatile flash memory—that is, memory that remembers even when the device is not powered.

On the right-hand edge of the board is the Serial Programming Connector. It offers another means of programming the Arduino without using the USB port. Because we do have a USB connection and software that makes it convenient to use, we will not avail ourselves of this feature.

In the top-left corner of the board next to the USB socket is the USB interface chip. This chip converts the signal levels used by the USB standard to levels that can be used directly by the Arduino board.

A Tour of a WiFi-Capable Arduino-Compatible

In contrast to the Arduino Uno, the board shown in Figure 1-2 is a low-cost Arduino-compatible board manufactured in China. This board has built-in WiFi, which is why we have chosen it for Chapter 10, the Internet of Things. In Chapter 6, we will meet this type of board again, along with some other types of Arduino-compatible boards.

Figure 1-2 *An Arduino-compatible Board (Lolin32 Lite).*

The board has most of the same things as an Arduino Uno. It has a USB connector, all though in this case it is a micro USB connector rather than the full-size connector of the Arduino Uno. It also has GPIO pins on its two long sides, and you generally have to solder your own headers onto these. You can either solder female header pins like the Uno, or more often, people solder on male header pins (which are usually supplied with the board). This board also has a battery connector for rechargeable LiPo battery in place of the DC barrel jack of the Uno. The board's microcontroller is labeled as a SoC (System on a Chip) to reflect the fact that it has built-in WiFi hardware rather than just being a simple microcontroller like the Uno uses.

The Origins of Arduino

Arduino was originally developed as an aid for teaching students. It was subsequently (in 2005) developed commercially by Massimo Banzi and David Cuartielles. It has since gone on to become enormously successful with makers, students, and artists for its ease of use and durability.

Another key factor in its success is that all the designs for Arduino are freely available under a Creative Commons license. This has allowed many lower-cost alternatives to the boards to appear. Only the name Arduino is protected, so such clones often have "*duino" names, such as Boarduino, Seeeduino, and Freeduino. Many big retailers sell only the official boards, which are nicely packaged and of high quality.

Yet another reason for the success of Arduino is that it is not limited to microcontroller boards. There are a huge number of Arduino-compatible shield boards that plug directly into the top of an Arduino board. Because shields are available for almost every conceivable application, you often can avoid using a soldering iron and instead plug together shields that can be stacked one upon another. The following are just a few of the most popular shields:

- Liquid Crystal Display (LCD) TFT displays
- Motor, which drives electric motors
- USB Host, which allows control of USB devices
- Relays, which switches relays from your Arduino

Figure 1-3 shows a motor shield (left) and relay shield (right).

Figure 1-3 *Motor and relay shields.*

Powering Up

When you buy an Arduino Uno board, it is usually preinstalled with a sample Blink program that will make the little built-in LED flash.

The LED marked *L* is wired up to one of the digital input output sockets on the board. It is connected to digital pin 13. This does not mean that pin 13 can only be used to light the LED; you can also use it as a normal digital input or output.

All you need to do to get your Arduino Uno up and running is supply it with some power. The easiest way to do this is to plug it into the USB port on your computer. You will need a type-A-to-type-B USB lead. This is the same type of lead that is normally used to connect a computer to a printer.

If everything is working OK, the LED should blink. New Arduino boards come with this Blink sketch already installed so that you can verify that the board works.

Installing the Software

To be able to install new sketches onto your Arduino board, you need to do more than supply power to it over the USB. You need to install the Arduino software (Figure 1-4).

Full and comprehensive instructions for installing this software on Windows, Linux, and Mac computers can be found at the Arduino website (www.arduino.cc).

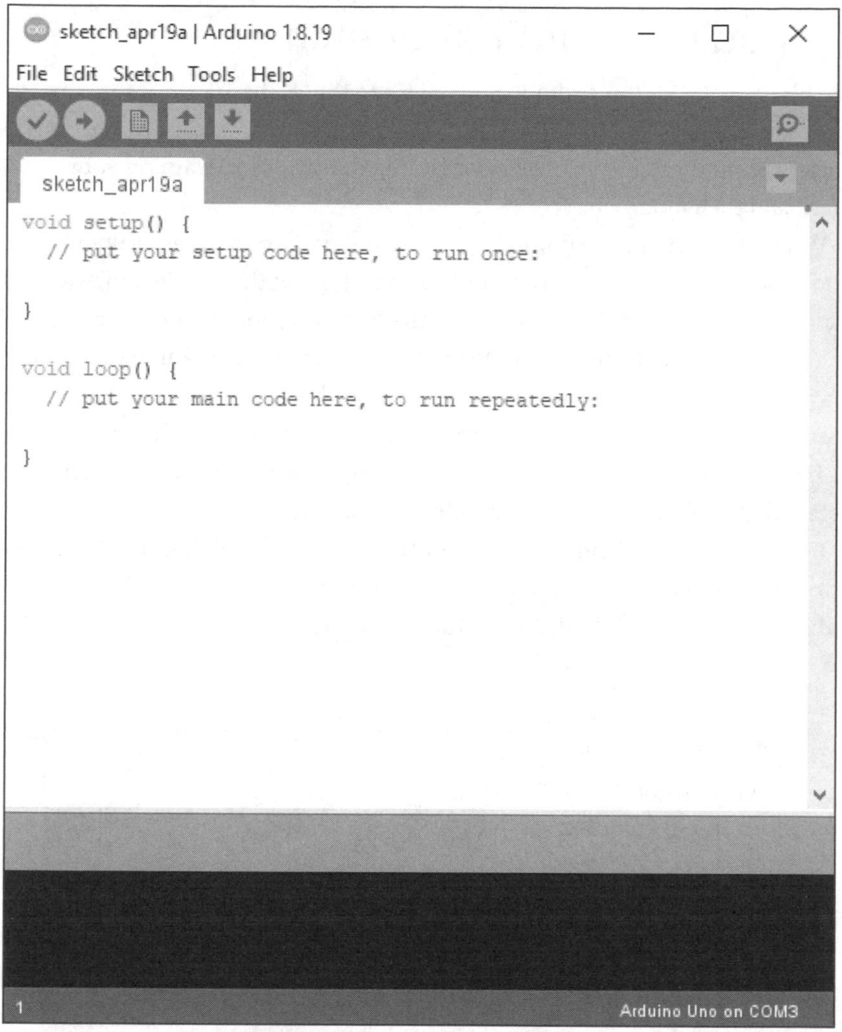

Figure 1-4   *The Arduino IDE application.*

Note that as well as the downloadable IDE that you run on your computer, there is also a web version. I recommend that you start with the downloadable IDE.

Once you have successfully installed the Arduino software and, depending on your platform, USB drivers, you should now be able to upload a program to the Arduino board.

Uploading Your First Sketch

The blinking LED is the Arduino equivalent to the "Hello World" program used in other languages as the traditional first program to run when learning a new language. Let's test out the environment by installing this program on your Arduino board and then modifying it.

When you start the Arduino application on your computer, it opens with an empty sketch. Fortunately, the application ships with a wide range of useful examples. So from the File menu, open the Blink example as shown in Figure 1-5.

You now need to transfer or upload that sketch to your Arduino board. So plug your Arduino board into your computer using the USB lead. You should see the green "On" LED on the Arduino light up. The Arduino board will probably already be flashing, as the boards are generally shipped with the Blink sketch already installed. But let's install it again and then modify it.

Before you can upload a sketch, you must tell the Arduino application what type of board you are using and which serial port you are connected to. Figures 1-6 and 1-7 show how you do this from the Tools menu.

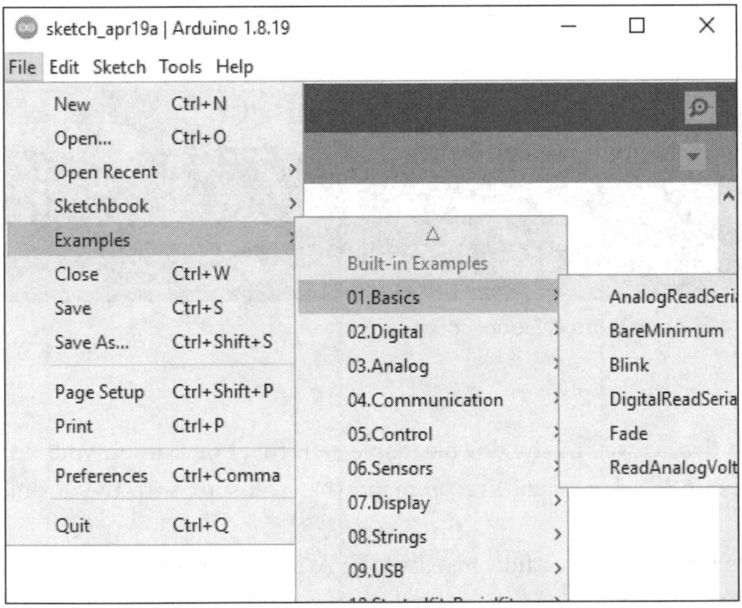

Figure 1-5   *The Blink sketch.*

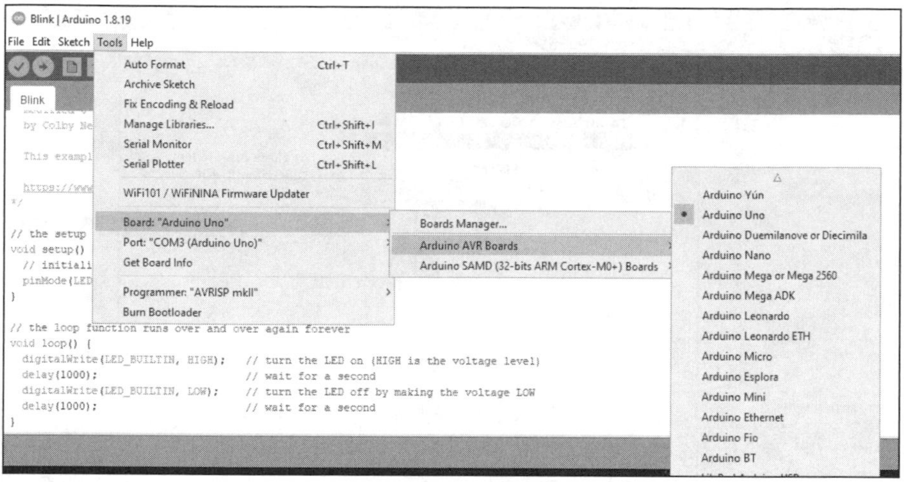

Figure 1-6   *Selecting the board type.*

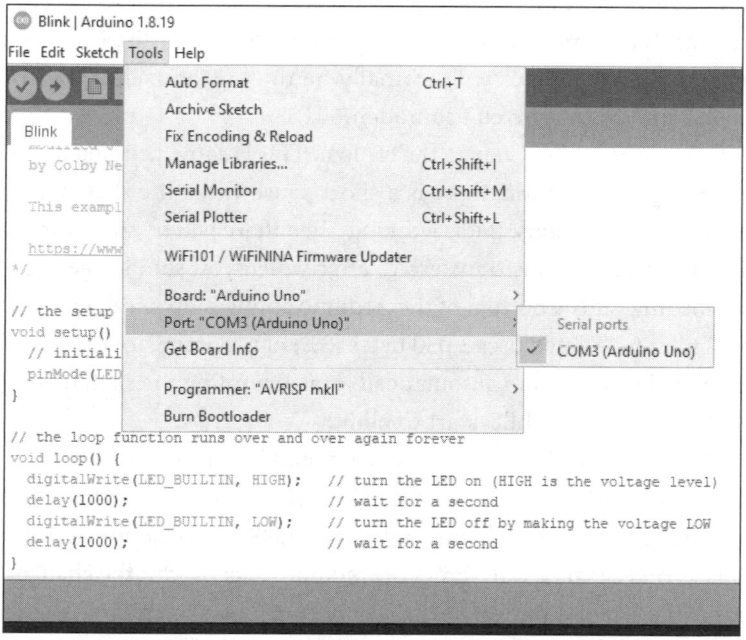

Figure 1-7   *Selecting the serial port (in Windows).*

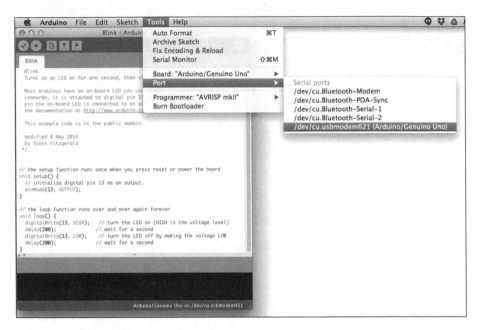

Figure 1-8 *Selecting the serial port (on a Mac).*

On a Windows machine, the serial port is always COM followed by a number. On Macs and Linux machines, you will see a much longer list of serial devices (see Figure 1-8). The device will normally be the bottom selection in the list, with a name similar to /dev/cu.usbmodem621.

Now click on the Upload icon in the toolbar. This is shown circled in Figure 1-9.

After you click the button, there is a short pause while the sketch is compiled and then the transfer begins. If it is working, then there will be some furious blinking of LEDs as the sketch is transferred, after which you should see the message "Done Uploading" at the bottom of the Arduino application window and a further message similar to "Sketch uses 1,030 bytes (3%) of program storage space."

Once uploaded, the board automatically starts running the sketch and you will see the yellow built-in "L" LED start to blink.

If this did not work, then check your serial and board type settings.

Now let's modify the sketch to make the LED blink faster. To do this, let's alter the two places in the sketch where there is a delay for 1,000 milliseconds so that the delay is 500 milliseconds. Figure 1-10 shows the modified sketch with the changes circled.

Click on the Upload button again. Then, once the sketch has uploaded, you should see your LED start to blink twice as fast as it did before.

Congratulations, you are now ready to start programming your Arduino. First, though, let's take a mini-tour of the Arduino application.

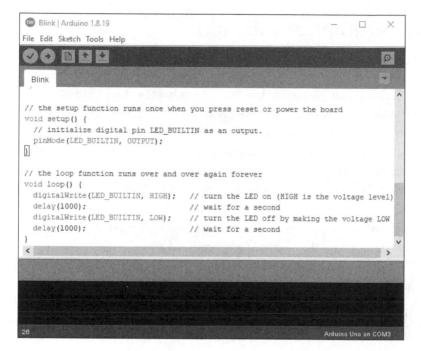

Figure 1-9 *Uploading the sketch.*

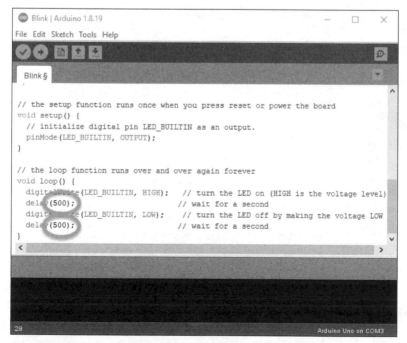

Figure 1-10 *Modifying the Blink sketch.*

The Arduino Application

Sketches in Arduino are like documents in a word processor. You can open them and copy parts from one to another. So you see options to Open, Save, and Save As in the File menu. You will not normally use Open because the Arduino application has the concept of a Sketchbook where all your sketches are kept carefully organized into folders. You gain access to the Sketchbook from the File menu. As you have just installed the Arduino application for the first time, your Sketchbook will be empty until you create some sketches.

As you have seen, the Arduino application comes with a selection of example sketches that can be very useful. Having modified the Blink example sketch, if you try and save it, you get a message that says, "Some files are marked read-only so you will need to save this sketch in a different location."

Try this now. Accept the default location, but change the filename to MyBlink, as shown in Figure 1-11.

Now if you go to the File menu and then click on Sketches, you will see MyBlink as one of the sketches listed. If you look at your computer's file system,

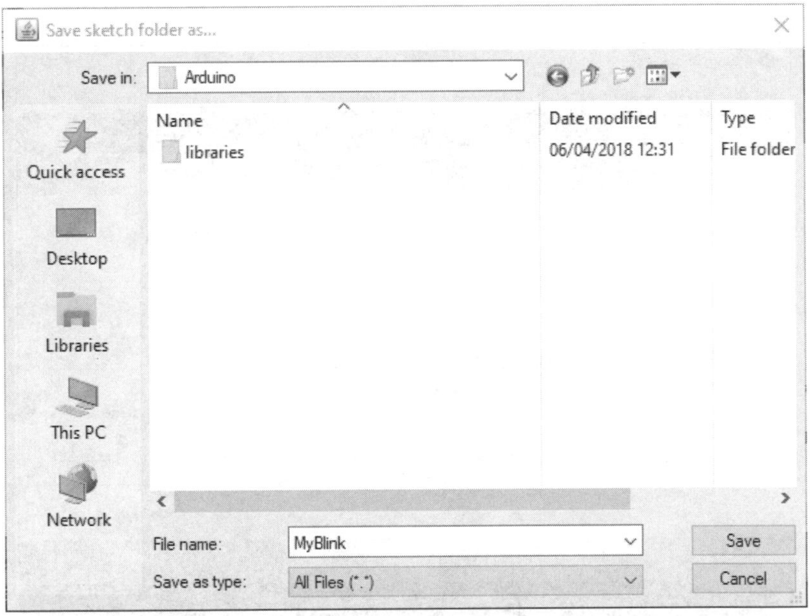

Figure 1-11 *Saving a copy of Blink.*

you will find that, on a PC, the sketch has been written into My Documents\ Arduino, and on Mac or Linux, it is in Documents/Arduino.

All of the sketches used in this book can be downloaded as a zip file from www.arduinobook.com. I suggest that now is the time to download this file and unzip it into the Arduino folder that contains the sketches. Figure 1-12 shows the files being extracted into the Arduino directory in Windows. In other words, when you have unzipped the folder, there should be two folders in your Arduino folder: one for the newly saved MyBlink and one called prog_arduino_3-main. The Programming Arduino folder will contain all the sketches, numbered according to chapter, so that sketch 02_01_blink, for example, is sketch 1 of Chapter 2.

These sketches will not appear in your Sketchbook menu until you quit the Arduino application and restart it. Do so now. Then your Sketchbook menu should look similar to that shown in Figure 1-13.

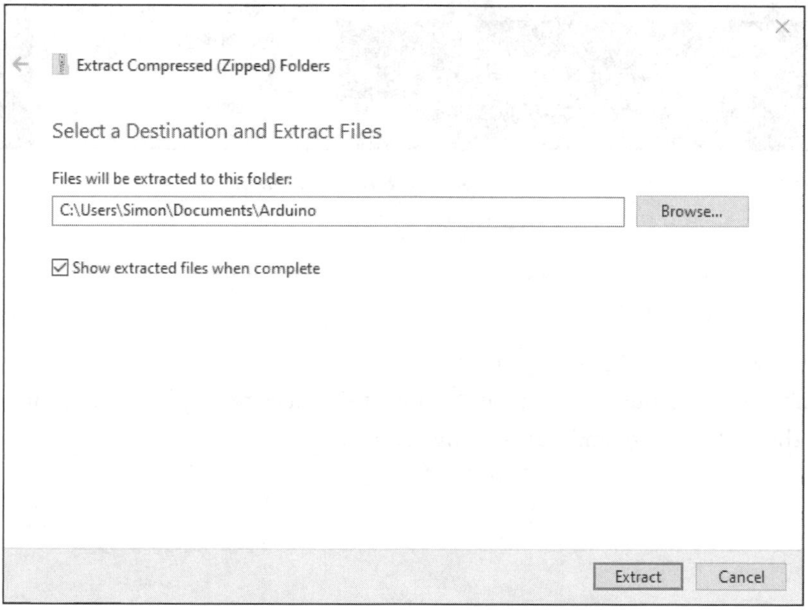

Figure 1-12 *Installing the sketches from the book.*

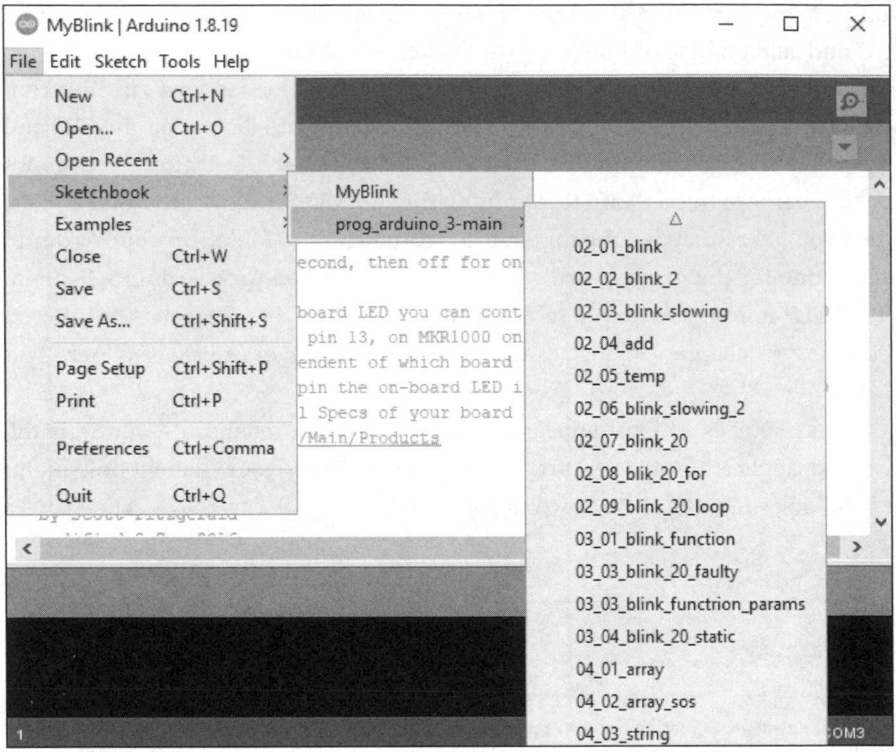

Figure 1-13   *Sketchbook with the book's sketches installed.*

Conclusion

Your environment is all set up and ready to go.

In the next chapter, we will look at some of the basic principles of the C language that the Arduino uses and start writing some code.

2

C Language Basics

The programming language used to program Arduinos is a language called C. In this chapter, you get to understand the basics of the C language. You will use what you learn here in every sketch you develop as an Arduino programmer. To get the most out of Arduino, you need to understand these fundamentals.

Programming

It is not uncommon for people to speak more than one language. In fact, the more you learn, the easier it seems to learn spoken languages as you start to find common patterns of grammar and vocabulary. The same is true of programming languages. So, if you have used any other programming language, you will quickly pick up C.

The good news is that the vocabulary of a programming language is far smaller than that of a spoken language, and because you write it rather than say it, the dictionary can always be at hand whenever you need to look things up. Also, the grammar and syntax of a programming language are extremely regular, and once you come to grips with a few simple concepts, learning more quickly becomes second nature.

It is best to think of a program—or a sketch, as programs are called in Arduino—as a list of instructions to be carried out in the order that they are written down. For example, suppose you were to write the following:

```
digitalWrite(13, HIGH);
delay(500);
digitalWrite(13, LOW);
```

These three lines would each do something. The first line would set the output of pin 13 to HIGH. This is the pin with a light-emitting diode (LED) built in to the Arduino Uno board, so at this point the LED would light. The second line would simply wait for 500 milliseconds (half a second) and then the third line would turn the LED back off again. So these three lines would achieve the goal of making the LED blink once for half a second.

You have already seen a bewildering array of punctuation used in strange ways and words that don't have spaces between them. A frustration of many new programmers is, "I know what I want to do, I just don't know what I need to write!" Fear not, all will be explained.

First of all, let's deal with the punctuation and the way the words are formed. These are both part of what is termed the syntax of the language. Most languages require you to be extremely precise about syntax, and one of the main rules is that names for things have to be a single word. That is, they cannot include spaces. So, **digitalWrite** is the name for something. It's the name of a built-in function (you'll learn more about functions later) that will do the job of setting an output pin on the Arduino board. Not only do you have to avoid spaces in names, but also names are case sensitive. So you must write **digitalWrite**, not **DigitalWrite** or **Digitalwrite.**

The function **digitalWrite** needs to know which pin to set and whether to set that pin HIGH or LOW. These two pieces of information are called *arguments*, which are said to be *passed* to a function when it is *called*. The arguments for a function must be enclosed in parentheses and separated by commas.

The convention is to place the opening parenthesis immediately after the last letter of the function's name and to put a space after the comma before the next argument. However, you can sprinkle space characters within the parentheses if you want.

If the function only has one argument, then there is no need for a comma.

Notice how each line ends with a semicolon. It would be more logical if they were periods, because the semicolon marks the end of one command, a bit like the end of a sentence.

In the next section, you will find out a bit more about what happens when you press the Upload button on the Arduino integrated development environment (IDE). Then you will be able to start trying out a few examples.

What Is a Programming Language?

It is perhaps a little surprising that we can get to Chapter 2 in a book about programming without defining exactly what a programming language is. We can recognize an Arduino sketch and probably have a rough idea of what it is trying to do, but we need to look a bit deeper into how some programming language code goes from being words on a page to something that does something real, like turn an LED on and off.

Figure 2-1 summarizes the process involved from typing code into the Arduino IDE to running the sketch on the board.

When you press the Upload button on your Arduino IDE, it launches a chain of events that results in your sketch being installed on the Arduino and being run. This is not as straightforward as simply taking the text that you typed into the editor and moving it to the Arduino board.

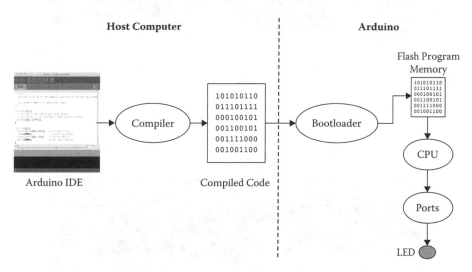

Figure 2-1 *From code to board.*

The first step is to do something called *compilation*. This takes the code you have written and translates it into machine code—the binary language that the Arduino understands. If you click the Verify button (leftmost check mark icon) on the Arduino IDE, this actually attempts to compile the C that you have written without trying to send the code to the Arduino IDE. A side-effect of compiling the code is that it is checked to make sure that it conforms to the rules of the C language.

If you type **Ciao Bella!** into your Arduino IDE and click on the Verify button, the results will be as shown in Figure 2-2.

The Arduino has tried to compile the words "Ciao Bella," and despite its Italian heritage, it has no idea what you are talking about. This text is not C. So, the result

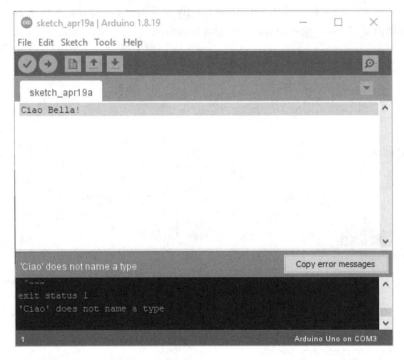

Figure 2-2 *Arduinos don't speak Italian.*

is that at the bottom of the screen we have that cryptic message "Ciao does not name a type." What this actually means is that there is a lot wrong with what you have written.

Let's try another example. This time we will try compiling a sketch with no code at all in it (see Figure 2-3).

This time, the compiler is telling you that your sketch does not have **setup** or **loop** functions. As you know from the Blink example that you ran in Chapter 1, you have to have some "boilerplate" code, as it is called, before you can add your own code into a sketch. In Arduino programming the "boilerplate" code takes the form of the "setup" and "loop" functions that must always be present in a sketch.

Figure 2-3 *No **setup** or **loop**.*

You will learn much more about functions later in the book, but for now, let's accept that you need this boilerplate code and just adapt your sketch so it will compile (see Figure 2-4).

The Arduino IDE has looked at your efforts at writing code and found them to be acceptable. It tells you this by saying "Done Compiling" and reporting the size of the sketch to you: 444 bytes. The IDE is also telling you that the maximum size is 32,256 bytes, so you still have lots of room to make your sketch bigger.

Let's examine this boilerplate code that will form the starting point for every sketch that you ever write. There are some new things here. For example, there is the word **void** and some curly braces. Let's deal with **void** first.

The line **void setup()** means that you are defining a function called **setup**. In Arduino, some functions are already defined for you, such as **digitalWrite** and **delay**, whereas you must or can define others for yourself. **setup** and **loop** are two functions that you must define for yourself in every sketch that you write.

Figure 2-4 *A sketch that will compile.*

The important thing to understand is that here you are not calling **setup** or **loop** like you would call **digitalWrite**, but you are actually creating these functions so that the Arduino system itself can call them. This is a difficult concept to grasp, but one way to think of it is as being similar to a definition in a legal document.

Most legal documents have a "definitions" section that might say, for example, something like the following:

```
Definitions.
The Author: The person or persons responsible for
creating the book
```

By defining a term in this way—for example, simply using the word "author" as shorthand for "The person or persons responsible for creating the book"—lawyers can make their documents shorter and more readable. Functions work much like such definitions. You define a function that you or the system itself can then use elsewhere in your sketches.

Going back to **void**, these two functions (**setup** and **loop**) do not return a value as some functions do, so you have to say that they are void, using the **void** keyword. If you imagine a function called **sin** that performed the trigonometric sine function, then this function would return a value. The value returned to use from the call would be the sine of the angle passed as its argument.

Rather like a legal definition uses words to define a term, we write functions in C that can then be called from C.

After the special keyword **void** comes the name of the function and then parentheses to contain any arguments. In this case, there are no arguments, but we still have to include the parentheses there. There is no semicolon after the closing parenthesis because we are defining a function rather than calling it, so we need to say what will happen when something does call the function.

Those things that are to happen when the function is called must be placed between curly braces. Curly braces and the code in between them are known as a *block* of code, and this is a concept that you will meet again later.

Note that although you do have to define both the functions **setup** and **loop**, you do not actually have to put any lines of code in them. However, failing to add code will make your sketch a little dull.

Blink—Again!

The reason that Arduino has the two functions **setup** and **loop** is to separate the things that only need to be done once, when the Arduino starts running its sketch, from the things that have to keep happening continuously.

The function **setup** will just be run once when the sketch starts. Let's add some code to it that will blink the LED built onto the board. Add the lines to your sketch so that it appears as follows and then upload them to your Arduino Uno.

```
void setup() {
  pinMode(13, OUTPUT);
  digitalWrite(13, HIGH);
}

void loop() {
}
```

You may have noticed that in the original Blink sketch LED_BUILTIN was used in preference to the pin number 13. LED_BUILTIN provides a way of making the code independent of the board. Although the built-in LED is always on pin 13 on an Arduino Uno, this is not the case for all Arduino and Arduino-compatible boards.

The **setup** function itself calls two built-in functions, **pinMode** and **digitalWrite**. You already know about **digitalWrite**, but **pinMode** is new. The function **pinMode** sets a particular pin to be either an input or an output. So, turning the LED on is actually a two-stage process. First, you have to set pin 13 to be an output, and second, you need to set that output to be high (5V).

When you run this sketch, on your board you will see that the L LED comes on and stays on. This is not very exciting, so let's at least try to make it flash by turning it on and off in the **loop** function rather than in the **setup** function.

You can leave the **pinMode** call in the **setup** function because you only need to call it once. The project would still work if you moved it into the loop, but there is no need and it is a good programming habit to do things only once if you only need to do them once. So modify your sketch so that it looks like this:

```
void setup() {
  pinMode(13, OUTPUT);
}

void loop() {
  digitalWrite(13, HIGH);
```

```
  delay(500);
  digitalWrite(13, LOW);
}
```

Run this sketch and see what happens. It may not be quite what you were expecting. The LED is basically on all the time. Hmm, why should this be?

Try stepping through the sketch a line at a time in your head:

1. Run **setup** and set pin 13 to be an output.

2. Run **loop** and set pin 13 to high (LED on).

3. Delay for half a second.

4. Set pin 13 to low (LED off).

5. Run **loop** again, going back to step 2, and set pin 13 to high (LED on).

The problem lies between steps 4 and 5. What is happening is that the LED is being turned off, but the very next thing that happens is that it gets turned on again. This happens so quickly that it appears that the LED is on all the time.

The microcontroller chip on the Arduino Uno can perform 16 million instructions per second. That's not 16 million C language commands, but it is still very fast. So, our LED will only be off for a few millionths of a second.

To fix the problem, you need to add another delay after you turn the LED off. Your code should now look like this:

```
// 02_01_blink
void setup() {
  pinMode(13, OUTPUT);
}

void loop() {
  digitalWrite(13, HIGH);
  delay(500);
  digitalWrite(13, LOW);
  delay(500);
}
```

Try again and your LED should blink away merrily once per second.

You may have noticed the comment at the top of the listing saying "sketch 02_01_blink." To save you some typing, we have uploaded to this book's website all the sketches with such a comment at the top. You can download them from www .arduinobook.com.

Variables

In this Blink example, you use pin 13 and have to refer to it in three places. If you decided to use a different pin, then you would have to change the code in three places. Similarly, if you wanted to change the rate of blinking, controlled by the argument to delay, you would have to change 500 to some other number in two places.

Variables can be thought of as giving a name to a number. Actually, they can be a lot more powerful than this, but for now, you will use them for this purpose.

When defining a variable in C, you have to specify the type of the variable. We want our variables to be whole numbers, which in C are called **int**s. So to define a variable called **ledPin** with a value of 13, you need to write the following:

```
int ledPin = 13;
```

Notice that because **ledPin** is a name, the same rules apply as those of function names. So, there cannot be any spaces. The convention is to start variables with a lowercase letter and begin each new word with an uppercase letter. Programmers will often call this "bumpy case" or "camel case."

Let's fit this into your Blink sketch as follows:

```
// 02_02_blink_2
int ledPin = 13;
int delayPeriod = 500;

void setup() {
  pinMode(ledPin, OUTPUT);
}

void loop() {
 digitalWrite(ledPin, HIGH);
 delay(delayPeriod);
 digitalWrite(ledPin, LOW);
 delay(delayPeriod);
}
```

We have also sneaked in another variable called **delayPeriod**.

Everywhere in the sketch where you used to refer to 13, you now refer to **ledPin**, and everywhere you used to refer to 500, you now refer to **delayPeriod**.

If you want to make the sketch blink faster, you can just change the value of **delayPeriod** in one place. Try changing it to 100 and running the sketch on your Arduino board.

There are other cunning things that you can do with variables. Let's modify your sketch so that the blinking starts really fast and gradually gets slower and slower, as if the Arduino is getting tired. To do this, all you need to do is to add something to the **delayPeriod** variable each time that you do a blink.

Modify the sketch by adding the single line at the end of the **loop** function so that it appears, as in the following listing, and then run the sketch on the Arduino board. Press the Reset button and see it start from a fast rate of flashing again.

```
// 02_02_blink_slowing
int ledPin = 13;
int delayPeriod = 100;

void setup() {
  pinMode(ledPin, OUTPUT);
}

void loop() {
 digitalWrite(ledPin, HIGH);
 delay(delayPeriod);
 digitalWrite(ledPin, LOW);
 delay(delayPeriod);
 delayPeriod = delayPeriod + 100;
}
```

Your Arduino is doing arithmetic now. Every time that **loop** is called, it will do the normal flash of the LED, but then it will add 100 to the variable **delayPeriod**. We will come back to arithmetic shortly, but first you need a better way than a flashing LED to see what the Arduino is up to.

Experiments in C

You need a way to test your experiments in C. One way is to put the C that you want to test out into the **setup** function, evaluate them on the Arduino, and then have the Arduino display any output back to something called the Serial Monitor, as shown in Figures 2-5 and 2-6.

The Serial Monitor is part of the Arduino IDE. You access it by clicking on the rightmost icon in the toolbar (it looks like a magnifying glass). Its purpose is to act as a communication channel between your computer and the Arduino. You can type a message in the text entry area at the top of the Serial

Figure 2-5 *Writing Output to the Serial Monitor.*

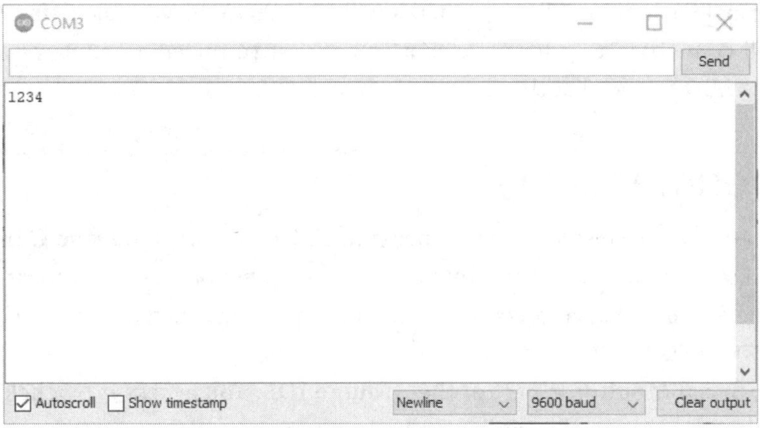

Figure 2-6 *The Serial Monitor.*

Monitor and when you press Return or click Send, it will send that message to the Arduino. Also if the Arduino has anything to say, this message will appear in the Serial Monitor. In both cases, the information is sent through the USB link.

As you would expect, there is a built-in function that you can use in your sketches to send a message back to the Serial Monitor. It is called **Serial.println** and it expects a single argument, which consists of the information that you want to send. This information is usually a variable.

You will use this mechanism to test out a few things that you can do with variables and arithmetic in C; frankly, it's the only way you can see the results of your experiments in C.

Numeric Variables and Arithmetic

The last thing you did was add the following line to your blinking sketch to increase the blinking period steadily:

```
delayPeriod = delayPeriod + 100;
```

Looking closely at this line, it consists of a variable name, then an equals sign, then what is called an expression (**delayPeriod + 100**). The equals sign does something called assignment. That is, it assigns a new value to a variable, and the value it is given is determined by what comes after the equals sign and before the semicolon. In this case, the new value to be given to the **delayPeriod** variable is the old value of **delayPeriod** plus 100.

Let's test out this new mechanism to see what the Arduino is up to by entering the following sketch, running it, and opening the Serial Monitor:

```
// 02_04_add
void setup() {
  Serial.begin(9600);
  int a = 2;
  int b = 2;
  int c = a + b;
  Serial.println(c);
}
void loop() {}
```

Figure 2-7 shows what you should see in the Serial Monitor after this code runs.

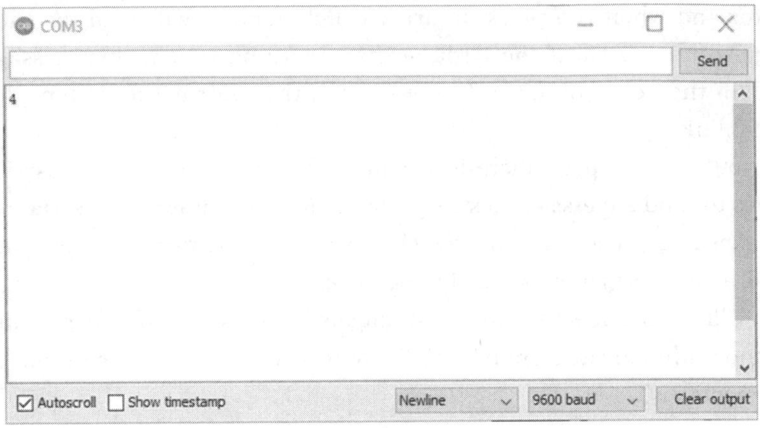

Figure 2-7   *Simple arithmetic.*

To take a slightly more complex example, the formula for converting a temperature in degrees Centigrade into degrees Fahrenheit is to multiply it by 9, divide by 5, and then add 32. So you could write that in a sketch like this:

```
// 02_05_temp
void setup() {
  Serial.begin(9600);
  int degC = 20;
  int degF;
  degF = degC * 9 / 5 + 32;
  Serial.println(degF);
}
void loop() {}
```

There are a few things to notice here. First, note the following line:

```
int degC = 20;
```

When we write such a line, we are actually doing two things: We are declaring an **int** variable called **degC**, and we are saying that its initial value will be 20. Alternatively, you could separate these two things and write the following:

```
int degC;
degC = 20;
```

You must declare any variable just once, essentially telling the compiler what type of variable it is—in this case, **int**. However, you can assign the variable a value as many times as you want:

```
int degC;
degC = 20;
degC = 30;
```

So, in the Centigrade to Fahrenheit example, you are defining the variable **degC** and giving it an initial value of 20, but when you define **degF**, it does not get an initial value. Its value gets assigned on the next line, according to the conversion formula, before being sent to the Serial Monitor for you to see.

Looking at the expression, you can see that you use the asterisk (*) for multiplication and the slash (/) for division. The arithmetic operators +, −, *, and / have an order of precedence—that is, multiplications are done first, then divisions, then additions and subtractions. This is in accordance with the usual use of arithmetic. However, sometimes it makes it clearer to use parentheses in the expressions. So, for example, you could write the following:

```
degF = ((degC * 9) / 5) + 32;
```

The expressions that you write can be as long and complex as you need them to be, and in addition to the usual arithmetic operators, there are other less commonly used operators and a big collection of various mathematical functions that are available to you. You will learn about these later.

Commands

The C language has a number of built-in commands. In this section, we explore some of these and see how they can be of use in your sketches.

if

In our sketches so far, we have assumed that your lines of programming will be executed in order one after the other, with no exceptions. But what if you don't want to do that? What if you only want to execute part of a sketch if some condition is true?

Let's return to our gradually slowing-down Blinking LED example. At the moment, it will gradually get slower and slower until each blink is lasting hours.

Let's look at how we can change it so that once it has slowed down to a certain point, it goes back to its fast starting speed.

To do this, you must use an **if** command; the modified sketch is as follows. Try it out.

```
// 02_06_blink_slowing_2
int ledPin = 13;
int delayPeriod = 100;

void setup() {
  pinMode(ledPin, OUTPUT);
}

void loop() {
  digitalWrite(ledPin, HIGH);
  delay(delayPeriod);
  digitalWrite(ledPin, LOW);
  delay(delayPeriod);
  delayPeriod = delayPeriod + 100;
  if (delayPeriod > 3000) {
    delayPeriod = 100;
  }
}
```

The **if** command looks a little like a function definition, but this resemblance is only superficial. The word in the parenthesis is not an argument; it is what is called *a condition*. So in this case, the condition is that the variable **delayPeriod** has a value that is greater than 3,000. If this is true, then the commands inside the curly braces will be executed. In this case, the code sets the value of **delayPeriod** back to 100.

If the condition is not true, then the Arduino will just continue on with the next thing. In this case, there is nothing after the "if," so the Arduino will run the **loop** function again.

Running through the sequence of events in your head will help you understand what is going on. So, here is what happens:

1. Arduino runs **setup** and initializes the LED pin to be an output.

2. Arduino starts running **loop**.

3. The LED turns on.

4. A delay occurs.

5. The LED turns off.

6. A delay occurs.

7. Add 100 to the **delayPeriod**.

8. If the delay period is greater than 3,000, set it back to 100.

9. Go back to step 3.

We used the symbol >, which means greater than. It is one example of what are called comparison operators. These operators are summarized in the following table:

Operator	Meaning	Examples	Result
<	Less than	**9 < 10**	true
		10 < 10	false
>	Greater than	**10 > 10**	false
		10 > 9	true
<=	Less than or equal to	**9 <= 10**	true
		10 <= 10	true
>=	Greater than or equal to	**10 >= 10**	true
		10 >= 9	true
==	Equal to	**9 == 9**	true
!=	Not equal to	**9 != 9**	false

To compare two numbers, you use the == command. This double equals sign is easily confused with the character =, which is used to assign values to variables.

There is another form of **if** that allows you to do one thing if the condition is true and another if it is false. We will use this in some practical examples later in the book.

for

In addition to executing different commands under different circumstances, you also often will want to run a series of commands a number of times in a program. You already know one way of doing this, using the **loop** function. As soon as all the commands in the **loop** function have been run, it will start again automatically. However, sometimes you need more control than that.

So, for example, let's say that you want to write a sketch that blinks 20 times, then pauses for 3 seconds, and then starts again. You could do that by just repeating the same code over and over again in your **loop** function, like this:

```
// 02_07_blink_20
int ledPin = 13;
int delayPeriod = 100;

void setup() {
  pinMode(ledPin, OUTPUT);
}

void loop() {
 digitalWrite(ledPin, HIGH);
 delay(delayPeriod);
 digitalWrite(ledPin, LOW);
 delay(delayPeriod);

 digitalWrite(ledPin, HIGH);
 delay(delayPeriod);
 digitalWrite(ledPin, LOW);
 delay(delayPeriod);

 digitalWrite(ledPin, HIGH);
 delay(delayPeriod);
 digitalWrite(ledPin, LOW);
 delay(delayPeriod);
// repeat the above 4 lines another 17 times

 delay(3000);
}
```

But this requires a lot of typing and there are several much better ways to do this. Let's start by looking at how you can use a **for** loop and then look at another way of doing it using a counter and an **if** statement.

The sketch to accomplish this with a **for** loop is, as you can see, a lot shorter and easier to maintain than the previous example:

```
// 02_08_blik_20_for
int ledPin = 13;
int delayPeriod = 100;
```

```
void setup() {
  pinMode(ledPin, OUTPUT);
}

void loop() {
  for (int i = 0; i < 20; i ++) {
    digitalWrite(ledPin, HIGH);
    delay(delayPeriod);
    digitalWrite(ledPin, LOW);
    delay(delayPeriod);
  }
 delay(3000);
}
```

The **for** loop looks a bit like a function that takes three arguments, although here those arguments are separated by semicolons rather than the usual commas. This is just a quirk of the C language. The compiler will soon tell you when you get it wrong.

The first thing in the parentheses after **for** is a variable declaration. This specifies a variable to be used as a counter variable and gives it an initial value—in this case, 0.

The second part is a condition that must be true for you to stay in the **loop**. In this case, you will stay in the **loop** as long as **i** is less than 20, but as soon as **i** is 20 or more, the program will stop doing the things inside the **loop**.

The final part is what to do every time you have done all the things in the **loop**. In this case, that is to increment **i** by 1 so that it can, after 20 trips around the **loop**, cease to be less than 20 and cause the program to exit the **loop**.

Try entering this code and running it. The only way to get familiar with the syntax and all that pesky punctuation is to type it in and have the compiler tell you when you have done something wrong. Eventually, it will all start to make sense.

One potential downside of this approach is that the **loop** function is going to take a long time. This is not a problem for this sketch, because all it is doing is flashing an LED. But often, the **loop** function in a sketch will also be checking that keys have been pressed or that serial communications have been received. If the processor is busy inside a **for** loop, it will not be able to do this. Generally, it is a good idea to make the **loop** function run as fast as possible so that it can be run as frequently as possible.

The following sketch shows how to achieve this:

```
// 02_09_blink_20_loop
int ledPin = 13;
int delayPeriod = 100;
int count = 0;
void setup() {
  pinMode(ledPin, OUTPUT);
}

void loop() {
 digitalWrite(ledPin, HIGH);
 delay(delayPeriod);
 digitalWrite(ledPin, LOW);
 delay(delayPeriod);
 count ++;
 if (count == 20) {
   count = 0;
   delay(3000);
 }
}
```

You may have noticed the following line:

```
count ++;
```

This is just C shorthand for the following:

```
count = count + 1;
```

So now each time that **loop** is run, it will take just a bit more than 200 milliseconds, unless it's the 20th time round the loop, in which case it will take the same plus the three seconds delay between each batch of 20 flashes. In fact, for some applications, even this is too slow, and purists would say that you should not use **delay** at all. The best solution depends on the application.

while

Another way of looping in C is to use the **while** command in place of the **for** command. You can accomplish the same thing as the preceding **for** example using a **while** command as follows:

```
int i = 0;
while (i < 20) {
  digitalWrite(ledPin, HIGH);
```

```
delay(delayPeriod);
digitalWrite(ledPin, LOW);
delay(delayPeriod);
i ++;
}
```

The expression in parentheses after **while** must be true to stay in the **loop**. When it is no longer true, then the sketch continues running the commands after the final curly brace.

Constants

For constant values like pin assignments that do not change during the running of a sketch, use the keyword **const**, which tells the compiler that the variable has a constant value and is not going to change.

As an example, you could define a pin assignment for a LED like this:

```
const int ledPin = 13;
```

Any sketch that you write will work just as well without the **const** keyword in front of any such variables, but it will make the program slightly smaller, something that can become significant as your sketches get bigger. In any case, it's a good habit to get into for any variables whose value is not going to change.

Arduino defines some of its own constants. For example, HIGH and LOW and OUTPUT are all constants, that actually represent numbers, but it's much easier to use names. Another constant that Arduino uses, that we could have used in our various blinking experiments is LED_BUILTIN. This has the value 13 for an Arduino Uno, but for other boards it may refer to a different pin number.

Conclusion

This chapter has got you started with C. You can make LEDs blink in various exciting ways and get the Arduino to send results back to you over the USB by using the **Serial.println** function. You also worked out how to use **if** and **for** commands to control the order in which your commands are executed, and learned a little about making an Arduino do some arithmetic.

In the next chapter, you will look more closely at functions. The chapter will also introduce the variable types other than the **int** type that you used in this chapter.

3

Functions

This chapter focuses mostly on the type of functions that you can write yourself rather than the built-in functions such as **digitalWrite** and **delay,** which are already defined for you.

The reason that you need to be able to write your own functions is that as sketches start to get a little complicated, then your **setup** and **loop** functions will grow and grow until they are long and complicated and it becomes difficult to see how they work.

The biggest problem in software development of any sort is managing complexity. The best programmers write software that is easy to look at and understand and requires very little in the way of explanation.

Functions are a key tool in creating easy-to-understand sketches that can be changed without difficulty or risk of the whole thing falling into a crumpled mess.

What Is a Function?

A function is a little like a program within a program. You can use it to wrap up some little thing that you want to do. A function that you define can be called from anywhere in your sketch and contains its own variables and its own list of commands. When the commands have been run, execution returns to the point just after wherever it was in the code that called the function.

By way of an example, code that flashes a light-emitting diode (LED) is a prime example of some code that should be put in a function. So let's modify our basic "blink 20 times" sketch to use a function that we will create called **flash,** shown next.

```
// 03_01_blink_function
const int ledPin = 13;
const int delayPeriod = 250;

void setup() {
  pinMode(ledPin, OUTPUT);
}

void loop() {
  for (int i = 0; i < 20; i ++) {
    flash();
  }
  delay(3000);
}

void flash() {
  digitalWrite(ledPin, HIGH);
  delay(delayPeriod);
  digitalWrite(ledPin, LOW);
  delay(delayPeriod);
}
```

So, all we have really done here is to move the four lines of code that flash the LED from the middle of the **for** loop to be in a function of their own called **flash**. Now you can make the LED flash any time you like by just calling the new function by writing **flash()**. Note the empty parentheses after the function name. This indicates that the function does not take any parameters. The delay value that it uses is set by the same **delayPeriod** variable that you used before.

Parameters

When dividing your sketch up into functions, it is often worth thinking about what service a function could provide. In the case of **flash**, this is fairly obvious. But this time, let's give this function parameters that tell it both how many times to flash and how short or long the flashes should be. Read through the following code and then I will explain just how parameters work in a little more detail.

```
// 03_02_blink_function_params
const int ledPin = 13;
const int delayPeriod = 250;
```

```
void setup() {
  pinMode(ledPin, OUTPUT);
}

void loop() {
  flash(20, delayPeriod);
  delay(3000);
}

void flash(int numFlashes, int d) {
  for (int i = 0; i < numFlashes; i ++) {
    digitalWrite(ledPin, HIGH);
    delay(d);
    digitalWrite(ledPin, LOW);
    delay(d);
  }
}
```

Now, if we look at our **loop** function, it has only two lines in it. We have moved the bulk of the work off to the **flash** function. Notice how when we call **flash** we now supply it with two arguments in parentheses.

Where we define the function at the bottom of the sketch, we have to declare the type of variable in the parameters. In this case, they are both **int**s. We are in fact defining new variables. However, these variables (**numFlashes** and **d**) can only be used within the **flash** function.

This is a good function because it wraps up everything you need in order to flash an LED. The only information that it needs from outside of the function is to which pin the LED is attached. If you wanted, you could make this a parameter too—something that would be well worth doing if you had more than one LED attached to your Arduino.

Global, Local, and Static Variables

As was mentioned before, parameters to a function can be used only inside that function. So, if you wrote the following code, you would get an error:

```
void indicate(int x) {
  flash(x, 10);
}
x = 15;
```

On the other hand, suppose you wrote this:

```
int x = 15;
void indicate(int x) {
  flash(x, 10);
}
```

This code would not result in a compilation error. However, you need to be careful, because you now actually have two variables called **x** and they can each have different values. The one that you declared on the first line is called a *global variable.* It is called *global* because it can be used anywhere you like in the program, including inside any functions.

However, because you use the same variable name **x** inside the function as a parameter, you cannot use the global variable **x** simply because whenever you refer to **x** inside the function, the "local" version of **x** has priority. The parameter **x** is said to shadow the global variable of the same name. This can lead to some confusion when trying to debug a project.

In addition to defining parameters, you can also define variables that are not parameters but are just for use within a function. These are called *local variables.* For example:

```
void indicate(int x) {
  int timesToFlash = x * 2;
  flash(timesToFlash, 10);
}
```

The local variable **timesToFlash** will only exist while the function is running. As soon as the function has finished its last command, it will disappear. This means that local variables are not accessible from anywhere in your program other than in the function in which they are defined.

So, for instance, the following example will cause an error:

```
void indicate(int x) {
  int timesToFlash = x * 2;
  flash(timesToFlash, 10);
}
timesToFlash = 15;
```

Seasoned programmers generally treat global variables with suspicion. The reason is that they go against the principal of encapsulation. The idea of *encapsulation* is that you should wrap up in a package everything that has to do with a particular feature. Hence functions are great for encapsulation. The problem with "globals" (as global variables are often called) is that they generally get defined at the beginning of a sketch and may then be used all over the sketch. Sometimes there is a perfectly legitimate reason for this. Other times, people use them in a lazy way when it would be far more appropriate to pass parameters. In our examples so far, **ledPin** is a good use of a global variable. It's also very convenient and easy to find up at the top of the sketch, making it easy to change.

Another feature of local variables is that their value is initialized every time the function is run. This is nowhere more true (and often inconvenient) than in the **loop** function of an Arduino sketch. Let's try and use a local variable in place of a global variable in one of the examples from the previous chapter:

```
// 03_03_blink_20_faulty
const int ledPin = 13;
const int delayPeriod = 250;

void setup() {
  pinMode(ledPin, OUTPUT);
}

void loop() {
 int count = 0;
 digitalWrite(ledPin, HIGH);
 delay(delayPeriod);
 digitalWrite(ledPin, LOW);
 delay(delayPeriod);
 count ++;
 if (count == 20) {
   count = 0;
   delay(3000);
 }
}
```

Sketch 03_03_blink_20_faulty is based on sketch 02_09_blink_20_loop, but attempts to use a local variable instead of the global variable to count the number of flashes.

This sketch is broken. It will not work, because every time **loop** is run, the variable **count** will be given the value 0 again, so **count** will never reach 20 and the LED will just keep flashing forever. The whole reason that we made **count** a global in the first place was so that its value would not be reset. The only place that we use **count** is in the **loop** function, so this is where it should be placed.

Fortunately, there is a mechanism in C that gets around this conundrum. It is the keyword **static**. When you use the keyword **static** in front of a variable declaration in a function, it has the effect of initializing the variable only the first time that the function is run. Perfect! That's just what is required in this situation. We can keep our variable in the function where it's used without it getting set back to 0 every time the function runs. Sketch 03_04_blink_20_static shows this in operation:

```
// 03_04_blink_20_static
const int ledPin = 13;
const int delayPeriod = 250;

void setup() {
  pinMode(ledPin, OUTPUT);
}

void loop() {
  static int count = 0;
  digitalWrite(ledPin, HIGH);
  delay(delayPeriod);
  digitalWrite(ledPin, LOW);
  delay(delayPeriod);
  count ++;
  if (count == 20) {
    count = 0;
    delay(3000);
  }
}
```

Return Values

Computer science, as an academic discipline, has as its parents mathematics and engineering. This heritage lingers on in many of the names associated with programming. The word *function* is itself a mathematical term. In mathematics, the input to the function (the argument or parameter) completely determines the output. We have written functions that take an input, but none that give us back a value. All our functions have been "void" functions. If a function returns a value, then you specify a return type.

Let's look at writing a function that takes a temperature in degrees Centigrade and returns the equivalent in degrees Fahrenheit:

```
int centToFaren(int c) {
  int f = c * 9 / 5 + 32;
  return f;
}
```

The function definition now starts with **int** rather than **void** to indicate that the function will return an **int** to whatever calls it. This might be a bit of code that looks like this:

```
int pleasantTemp = centToFaren(20);
```

Any non-void function has to have a **return** statement in it. If you do not put one in, the compiler will tell you that it is missing. You can have more than one **return** in the same function. This might arise if you have an **if** statement with alternative actions based on some condition. Some programmers frown on this, but if your functions are small (as all functions should be), then this practice will not be a problem.

The value after **return** can be an expression; it does not have to just be the name of a variable. So you could compress the preceding example into the following:

```
int centToFaren(int c) {
  return (c * 9 / 5 + 32);
}
```

If the expression being returned is more than just a variable name, then it should be enclosed in parentheses as in the preceding example.

Other Variable Types

All our examples of variables so far have been **int** variables. This is by far the most commonly used variable type, but there are some others that you should be aware of.

Floats

One such type, which is relevant to the previous temperature conversion example, is **float**. This variable type represents floating point numbers—that is, numbers that may have a decimal point in them, such as 1.23. You need this variable type when whole numbers are just not precise enough.

Note the following formula:

```
f = c * 9 / 5 + 32
```

If you give **c** the value 17, then **f** will be 17 * 9 / 5 + 32 or 62.6. But if **f** is an **int**, then the value will be truncated to 62.

The problem becomes even worse if we are not careful of the order in which we evaluate things. For instance, suppose that we did the division first, as follows:

```
f = (c / 5) * 9 + 32
```

Then in normal math terms, the result would still be 62.6, but if all the numbers are **int**s, then the calculation would proceed as follows:

1. 17 is divided by 5, which gives 3.4, which is then truncated to 3.

2. 3 is then multiplied by 9 and 32 is added to give a result of 59—which is quite a long way from 62.6.

For circumstances like this, we can use **floats**. In the following example, our temperature conversion function is rewritten to use **floats**:

```
float centToFaren(float c) {
  float f = c * 9.0 / 5.0 + 32.0;
  return f;
}
```

Notice how we have added .0 to the end of our constants. This ensures that the compiler knows to treat them as **floats** rather than **ints**.

Boolean

Boolean values are logical. They have a value that is either true or false.

In the C language, *Boolean* is spelled with a lowercase *b*, but in general use, *Boolean* has an uppercase initial letter, as it is named after the mathematician George Boole, who invented the Boolean logic that is crucial to computer science.

You may not realize it, but you have already met Boolean values when we were looking at the **if** command. The condition in an **if** statement, such as (**count == 20**), is actually an expression that yields a **boolean** result. The operator **==** is called a comparison operator. Whereas + is an arithmetic operator that adds two numbers together, == is a comparison operator that compares two numbers and returns a value of either **true** or **false**.

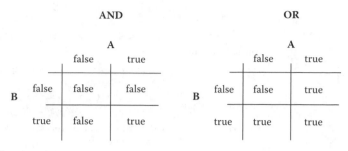

Figure 3-1 *Truth tables.*

You can define Boolean variables and use them as follows:

```
boolean tooBig = (x > 10);
if (tooBig) {
  x = 5;
}
```

Boolean values can be manipulated using Boolean operators. So, similar to how you can perform arithmetic on numbers, you can also perform operations on Boolean values. The most commonly used Boolean operators are **and**, which is written as **&&**, and **or**, which is written as ||.

Figure 3-1 shows truth tables for the **and** and **or** operators.

From the truth tables in Figure 3-1, you can see that for **and**, if both A and B are true, then the result will be true; otherwise, the result will be false.

On the other hand, with the **or** operator, if either A or B or both A and B are true, then the result will be true. The result will be false only if neither A nor B is true.

In addition to **and** and **or**, there is the **not** operator, written as **!**. You will not be surprised to learn that "not true" is false and "not false" is true.

You can combine these operators into Boolean expressions in your **if** statements, as the following example illustrates:

```
if ((x > 10) && (x < 50))
```

Other Data Types

As you have seen, the **int** and occasionally the **float** data types are fine for most situations; however, some other numeric types can be useful under some circumstances. In an Arduino sketch, the **int** type uses 16 bits (binary digits). This allows it to represent numbers between −32768 and 32767.

Type	Memory (Bytes)	Range	Notes
boolean	1	True or false (0 or 1)	
char	1	−128 to +127	Used to represent an American Standard Code for Information Interchange (ASCII) character code; e.g., *A* is represented as 65. Its negative numbers are not normally used.
byte	1	0 to 255	Often used for communicating serial data, as a single unit of data.
int	2	−32768 to +32767	
unsigned int	2	0 to 65535	Can be used for extra precision where negative numbers are not needed. Use with caution, as arithmetic with **int**s may cause unexpected results.
long	4	−2,147,483,648 to 2,147,483,647	Needed only for representing very big numbers.
unsigned long	4	0 to 4,294,967,295	See **unsigned int**.
float	4	−3.4028235E+38 to + 3.4028235E+38	
double	4	Same as **float**	Normally, this would be 8 bytes and higher precision than **float** with a greater range. However, on Arduino, it is the same as **float**.

Table 3-1 *Data Types in C*

Other data types available to you are summarized in Table 3-1. This table is provided mainly for reference. You will use some of these other types as you progress through the book. Note that Arduino devices using 32-bit architectures, such as the ESP32 boards, have 4-byte **int**s, giving them the same range as **long**s on an Arduino Uno.

One thing to consider is that if data types exceed their range, then strange things happen. So, if you have a byte variable with 255 in it and you add 1 to it, you get 0. More alarmingly, if you have an **int** variable with 32767 and you add 1 to it, you will end up with −32768.

Until you are completely comfortable with these different data types, I would recommend sticking to **int**, as it works for pretty much everything.

Coding Style

The C compiler does not really care about how you lay out your code. For all it cares, you can write everything on a single line with semicolons between each statement. However, well-laid-out, neat code is much easier to read and maintain than poorly laid-out code. In this sense, reading code is just like reading a book: Formatting is important.

To some extent, formatting is a matter of personal taste. No one likes to think that he has bad taste, so arguments about how code should look can become personal. It is not unknown for programmers, on being required to do something with someone else's code, to start by reformatting all the code into their preferred style of presentation.

As an answer to this problem, coding standards are often laid down to encourage everyone to lay out his or her code in the same way and adopt "good practice" when writing programs.

The C language has a de facto standard that has evolved over the years, and this book is generally faithful to that standard.

Indentation

In the example sketches that you have seen, you can see that we often indent the program code from the left margin. So, for example when defining a **void** function, the **void** keyword is at the left and then all the text within the curly braces is indented. The amount of indentation does not really matter. Some people use two spaces, some four. You can also press TAB to indent. In this book, we use two spaces for indentation.

If you were to have an **if** statement inside a function definition, then once again you would add two more spaces for the lines within the curly braces of the **if** command, as in the following example:

```c
void loop() {
  static int count = 0;
  count ++;
  if (count == 20) {
    count = 0;
    delay(3000);
  }
}
```

You might include another **if** inside the first **if**, which would add yet another level of indentation, making six spaces from the left margin.

All of this might sound a bit trivial, but if you ever sort through someone else's badly formatted sketches, you will find it very difficult.

Opening Braces

There are two schools of thought as to where to put the first curly brace in a function definition, **if** statement, or **for** loop. One way is to place the curly brace on the line after the rest of the command, as shown below, or put it on the same line, as we have in all the examples so far.

```
void loop()
{
  static int count = 0;
  count ++;
  if (count ==20)
  {
    count = 0;
    delay(3000);
  }
}
```

The style we use in this book is most commonly used in the Java programming language, which shares much of the same syntax as C.

Whitespace

The compiler ignores spaces, tabs, and new lines, apart from using them as a way of separating the "tokens" or words in your sketch. Thus the following example will compile without a problem:

```
void loop() {static int
count=0;count++;if(
count==20){count=0;
delay(3000);}}
```

This will work, but good luck trying to read it.

Where assignments are made, some people will write the following:

```
int a = 10;
```

But others will write the following:

```
int a=10;
```

Which of these two styles you use really does not matter, but it is a good idea to be consistent. I use the first form.

Comments

Comments are text that is kept in your sketch along with all the real program code, but which actually performs no programming function whatsoever. The sole purpose of comments is to be a reminder to you or others as to why the code is written as it is. A comment line may also be used to present a title.

The compiler will completely ignore any text that is marked as being a comment. We have included comments as titles at the top of many of the sketches in the book so far.

There are two forms of syntax for comments:

- The single line comment that starts with // and finishes at the end of the line
- The multiline comment that starts with a /* and ends with a */

The following is an example using both forms of comments:

```
/* A not very useful loop function.
Written by: Simon Monk
To illustrate the concept of comments
*/
void loop() {
  static int count = 0;
  count ++; // a single line comment
  if (count == 20) {
    count = 0;
    delay(3000);
  }
}
```

In this book, I mostly stick to the single-line comment format.

Good comments help explain what is happening in a sketch or how to use the sketch. They are useful if others are going to use your sketch, but equally useful to yourself when you are looking at a sketch that you have not worked on for a few weeks.

Some people are told in programming courses that the more comments, the better. Most seasoned programmers will tell you that well-written code requires very little in the way of comments because it is self-explanatory. You should use comments for the following reasons:

- To explain anything you have done that is a little tricky or not immediately obvious

- To describe anything that the user needs to do that is not part of the program; for example, // **this pin should be connected to the transistor controlling the relay**

- To leave yourself notes; for example, // **todo: tidy this - it's a mess**

This last point illustrates a useful technique of **todo**s in comments. Programmers often put **todo**s in their code to remind themselves of something they need to do later. They can always use the search facility in the Arduino IDE to find all occurrences of // **todo** in their sketch.

The following are *not* good examples of reasons you should use comments:

- To state the blatantly obvious; for example, **a = a + 1; // add 1 to a**.

- To explain badly written code. Don't comment on it; just write it clearly in the first place.

Conclusion

This has been a bit of a theoretical chapter. You have had to absorb some new abstract concepts concerned with organizing our sketches into functions and adopting a style of programming that will save you time in the long run.

In the next chapter, you can start to apply some of what you have learned and look at better ways of structuring your data and using text strings.

4

Arrays and Strings

After reading Chapter 3, you have a reasonable appreciation as to how to structure your sketches to make your life easier. If there is one thing that a good programmer likes, it's an easy life. Now our attention is going to turn to the data that you use in your sketches.

The book *Algorithms + Data Structures = Programs* by Niklaus Wirth (Prentice-Hall, 1976) has been around for a good while now, but still manages to capture the essences of computer science and programming in particular. I can strongly recommend it to anyone who finds themselves bitten by the programming bug. It also captures the idea that to write a good program, you need to think about both the algorithm (what you do) and the structure of the data you use.

You have looked at **loop**s, **if** statements, and what is called the "algorithmic" side of programming an Arduino; you are now going to turn to how you structure your data.

Arrays

Arrays are a way of containing a list of values. The variables that you have met so far have contained only a single value, usually an **int**. By contrast, an array contains a list of values, and you can access any one of those values by its position in the list.

C, in common with the majority of programming languages, begins its index positions at 0 rather than 1. This means that the first element is actually element zero.

To illustrate the use of arrays, we could create an example application that repeatedly flashes "SOS" in Morse code using the Arduino board's built-in LED.

Morse code used to be a vital method of communication in the 19th and 20th centuries. Because of its coding of letters as a series of long and short dots, Morse code can be sent over telegraph wires, over a radio link, and using signaling lights. The letters "SOS" (often taken to mean "save our souls") are still recognized as an international signal of distress.

The letter "S" is represented as three short flashes (dots) and the letter "O" by three long flashes (dashes). You are going to use an array of **int**s to hold the duration of each flash that you are going to make. You can then use a **for** loop to step through each of the items in the array, making a flash of the appropriate duration.

First, let's have a look at how you are going to create an array of **int**s containing the durations.

```
int durations[] = {200, 200, 200, 500, 500, 500, 200, 200, 200};
```

You indicate that a variable contains an array by placing [] after the variable name.

In this case, you are going to set the values for the durations at the time that you create the array. The syntax for doing this is to use curly braces and then values each separated by commas. Don't forget the semicolon on the end of the line.

You can access any given element of the array using the square bracket notation. So, if you want to get the first element of the array, you can write the following:

```
durations[0]
```

To illustrate this, sketch 04_01_array will create an array and then print out all its values to the Serial Monitor:

```
// 04_01_array

int durations[] = {200, 200, 200, 500, 500, 500, 200, 200, 200};

void setup() {
  Serial.begin(9600);
  for (int i = 0; i < 9; i++) {
    Serial.println(durations[i]);
  }
}

void loop() {}
```

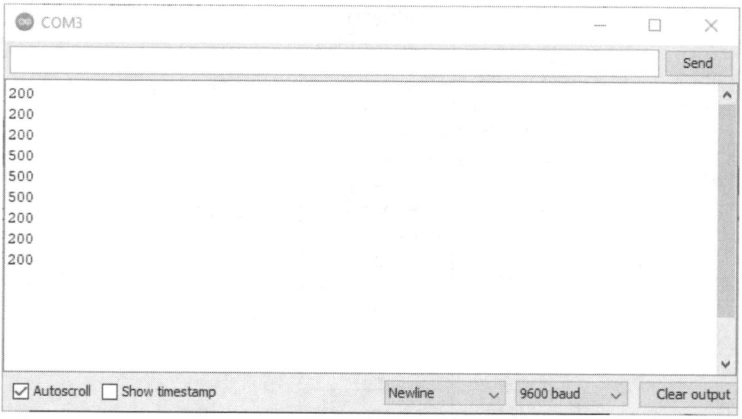

Figure 4-1   *The Serial Monitor showing the output of sketch 04_01_array.*

Note that you can use the keyword **const** with arrays as well as ordinary variables, just as long as you do not intend to modify the array within your sketch.

Upload the sketch to your board and then open the Serial Monitor. If all is well, you will see something like Figure 4-1.

This is quite neat, because if you wanted to add more durations to the array, all you would need to do is add them to the list inside the curly braces and change "9" in the **for** loop to the new size of the array.

You have to be a little careful with arrays, because the compiler will not try and stop you from accessing elements of data that are beyond the end of the array. This is because the array is really a pointer to an address in memory, as shown in Figure 4-2.

Programs keep their data, both ordinary variables and arrays, in *memory*. Computer memory is arranged much more rigidly than the human kind of memory. It is easiest to think of the memory in an Arduino as a collection of pigeonholes. When you define an array of nine elements, for example, the next available nine pigeonholes are reserved for its use and the variable is said to point at the first pigeonhole or *element* of the array.

Going back to our point about access being allowed beyond the bounds of your array, if you decided to access **durations[10]**, then you would still get back an **int**, but the value of this **int** could be anything. This is in itself fairly harmless, except that if you accidentally get a value outside of the array, you are likely to get confusing results in your sketch.

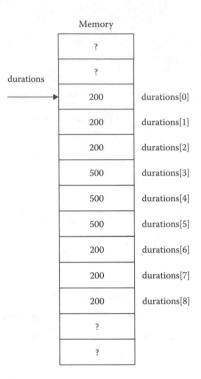

Figure 4-2 *Arrays and pointers.*

However, what is far worse is if you try changing a value outside of the size of the array. For instance, if you were to include something like the following in your program, the results could simply make your sketch break:

```
durations[10] = 0;
```

The pigeonhole **durations[10]** may be in use as some completely different variable. So always make sure that you do not go outside of the size of the array. If your sketch starts behaving strangely, then check for this kind of problem.

Morse Code SOS Using Arrays

Sketch 04_02_array_sos shows how you can use an array to make your emergency signal of SOS:

```
// 04_02_array_sos
const int ledPin = 13;

int durations[] = {200, 200, 200, 500, 500, 500, 200, 200, 200};
```

```
void setup() {
  pinMode(ledPin, OUTPUT);
}

void loop() {
  for (int i = 0; i < 9; i++) {
    flash(durations[i]);
  }
  delay(1000);
}

void flash(int delayPeriod) {
  digitalWrite(ledPin, HIGH);
  delay(delayPeriod);
  digitalWrite(ledPin, LOW);
  delay(delayPeriod);
}
```

An obvious advantage of this approach is that it is very easy to change the message by simply altering the **durations** array. In sketch 04_05_morse_flasher, you will take the use of arrays a stage further to make a more general-purpose Morse code flasher.

String Arrays

In the programming world, the word *string* has nothing to do with long thin stuff that you tie knots in. A string is a sequence of characters. It's the way you can get your Arduino to deal with text. For example, sketch 04_03_string will repeatedly send the text "Hello" to the Serial Monitor one time per second:

```
// 04_03_string
void setup() {
  Serial.begin(9600);
}

void loop() {
  Serial.println("Hello");
  delay(1000);
}
```

String Literals

String literals are enclosed in double quotation marks. They are literal in the sense that the string is a constant, rather like the **int** 123.

Character	ASCII Code (Decimal)
a–z	97–122
A–Z	65–90
0–9	48–57
space	32

Table 4-1 *Common ASCII Codes*

As you would expect, you can put strings in a variable. There is also an advanced string library, but for now you will use standard C strings, such as the one in sketch 04_03_string.

In C, a string literal is actually an array of the type **char**. The type **char** is a bit like **int** in that it is a number, but that number is between 0 and 127 and represents one character. The character may be a letter of the alphabet, a number, a punctuation mark, or a special character such as a tab or a line feed. These number codes for letters use a standard called ASCII. Some of the most commonly used ASCII codes are shown in Table 4-1.

The string literal "Hello" is actually an array of characters, as shown in Figure 4-3.

Note that the string literal has a special null character (\0) at the end. This character is used to indicate the end of the string.

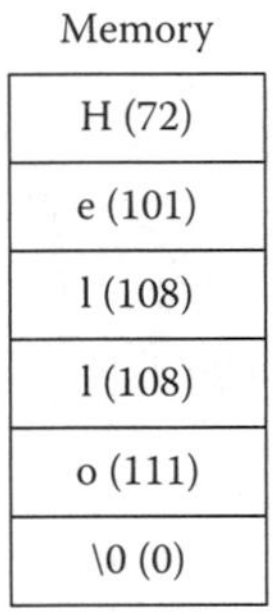

Memory

H (72)
e (101)
l (108)
l (108)
o (111)
\0 (0)

Figure 4-3 *The string literal "Hello".*

String Variables

As you would expect, string variables are very similar to array variables, except that there is a useful shorthand method for defining their initial value.

```
char name[] = "Hello";
```

This defines an array of characters and initializes it to the word "Hello." It will also add a final null value (ASCII 0) to mark the end of the string.

Although the preceding example is most consistent with what you know about writing arrays, it would be more common to write the following:

```
char *name = "Hello";
```

This is equivalent, and the * indicates a pointer. The idea is that **name** points to the first **char** element of the **char** array. That is the memory location that contains the letter *H*.

You can rewrite sketch 04_03_string to use a variable as well as a string constant, as follows:

```
// 04_04_string_var
char message[] = "Hello";

void setup() {
  Serial.begin(9600);
}

void loop() {
  Serial.println(message);
  delay(1000);
}
```

A Morse Code Translator

Let's put together what you have learned about arrays and strings to build a more complex sketch that will accept any message from the Serial Monitor and flash it out on the built-in LED.

The letters in Morse code are shown in Table 4-2.

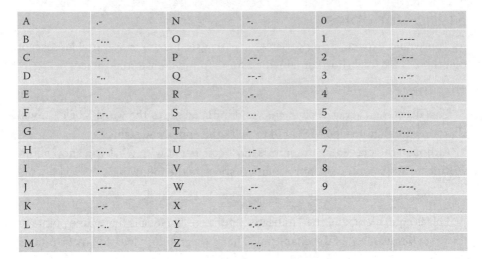

A	.-	N	-.	0	-----
B	-...	O	---	1	.----
C	-.-.	P	.--.	2	..---
D	-..	Q	--.-	3	...--
E	.	R	.-.	4	....-
F	..-.	S	...	5	.....
G	--.	T	-	6	-....
H	....	U	..-	7	--...
I	..	V	...-	8	---..
J	.---	W	.--	9	----.
K	-.-	X	-..-		
L	.-..	Y	-.--		
M	--	Z	--..		

Table 4-2 *Morse Code Letters*

Some of the rules of Morse code are that a dash is three times as long as a dot, the time between each dash or dot is equal to the duration of a dot, the space between two letters is the same length as a dash, and the space between two words is the same duration as seven dots.

For this project, we will not worry about punctuation, although it would be an interesting exercise for you to try adding this to the sketch. For a full list of all the Morse characters, see en.wikipedia.org/wiki/Morse_code.

Data

You are going to build this example a step at a time, starting with the data structure that you are going to use to represent the codes.

It is important to understand that there is no one solution to this problem. Different programmers will come up with different ways to solve it. So, it is a mistake to think to yourself, "I would never have come up with that." Well, no, quite possibly you would come up with something different and better. Everyone thinks in different ways, and this solution happens to be the one that first popped into the author's head.

Representing the data is all about finding a way of expressing Table 4-2 in C. In fact, you are going to split the data into two tables: one for the letters and one for the numbers. The data structure for the letters is as follows:

```
char* letters[] = {
    ".-", "-...", "-.-.", "-..", ".", // A-I
    "..-.", "--.", "....", "..",
    ".---", "-.-", ".-..", "--", "-.", // J-R
    "---", ".--", "--.-", ".-.",
    "...", "-", "..-", "...-", ".--", // S-Z
    "-..-", "-.--", "--.." 
};
```

What you have here is an array of string literals. So, because a string literal is actually an array of **char**, what you actually have here is an array of arrays—something that is perfectly legal and really quite useful.

This means that to find Morse for *A*, you would access **letters[0]**, which would give you the string **.-**. This approach is not terribly efficient, because you are using a whole byte (8 bits) of memory to represent a dash or a dot, which could be represented in a bit. However, you can easily justify this approach by saying that the total number of bytes is still only about 90 and we do have 2,048 bytes to play with. Equally importantly, it makes the code easy to understand.

Numbers use the same approach:

```
char* numbers[] = {
    "-----", ".----", "..---", "...--", "....-",
    ".....", "-....", "--...", "---..", "----." };
```

Globals and Setup

You need to define a couple of global variables: one for the delay period for a dot, and one to define which pin the LED is attached to:

```
const int dotDelay = 200;
const int ledPin = 13;
```

The **setup** function is pretty simple; you just need to set the **ledPin** as an output and set up the serial port:

```
void setup() {
  pinMode(ledPin, OUTPUT);
  Serial.begin(9600);
}
```

The loop Function

You are now going to start on the real processing work in the **loop** function. The algorithm for this function is as follows:

- If there is a character to read from USB:
  - If it's a letter, flash it using the letters array.
  - If it's a number, flash it using the numbers array.
  - If it's a space, flash four times the dot delay.

That's all. You should not think too far ahead. This algorithm represents what you want to do, or what your *intention* is, and this style of programming is called *programming by intention*.

If you write this algorithm in C, it will look like this:

```
void loop() {
  char ch;
  if (Serial.available() > 0) {
    ch = Serial.read();
    if (ch >= 'a' && ch <= 'z') {
      flashSequence(letters[ch - 'a']);
    }
    else if (ch >= 'A' && ch <= 'Z') {
      flashSequence(letters[ch - 'A']);
    }
    else if (ch >= '0' && ch <= '9') {
      flashSequence(numbers[ch - '0']);
    }
    else if (ch == ' ') {
      delay(dotDelay * 4);        // gap between words
    }
  }
}
```

There are a few things here that need explaining. First, there is **Serial.available()**. To understand this, you first need to know a little about how an Arduino communicates with your computer over USB. Figure 4-4 summarizes this process.

In the situation where the computer is sending data from the Serial Monitor to the Arduino board, then the USB is converted from the USB signal levels and protocol to something that the microcontroller on the Arduino board can use. This conversion happens in a special-purpose chip on the Arduino board. The

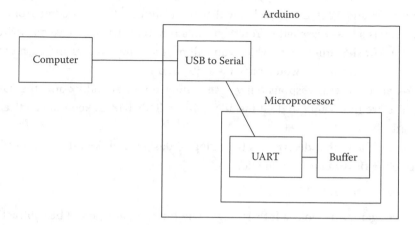

Figure 4-4 *Serial communication with an Arduino.*

data is then received by a part of the microcontroller called the Universal Asynchronous Receiver/Transmitter (UART). The UART places the data it receives into a buffer. The buffer is a special area of memory (128 bytes) that can hold data that is removed as soon as it is read.

This communication happens regardless of what your sketch is doing. So, even though you may be merrily flashing LEDs, data will still arrive in the buffer and sit there until you are ready to read it. You can think of the buffer as being a bit like an e-mail inbox.

The way that you check to see whether you "have mail" is to use the function **Serial.available()**. This function returns the number of bytes of data in the buffer that are waiting for you to read. If there are no messages waiting to be read, then the function returns 0. This is why the **if** statement checks to see that there are more than zero bytes available to read, and if they are, then the first thing that the statement does is read the next available **char**, using the function **Serial.read()**. This function gets assigned to the local variable **ch**.

Next is another **if** to decide what kind of thing it is that you want to flash:

```
if (ch >= 'a' && ch <= 'z') {
    flashSequence(letters[ch - 'a']);
}
```

At first, this might seem a bit strange. You are using <= and >= to compare characters. You can do that because each character is actually represented by a number (its ASCII code). So, if the code for the character is somewhere between *a* and *z* (97 and 122), then you know that the character that has come from the computer is a lowercase letter. You then call a function that you have not written yet called **flashSequence**, to which you will pass a string of dots and dashes; for example, to flash *a*, you would pass it .- as its argument.

You are devolving responsibility to this function for actually doing the flashing. You are not trying to do it inside the **loop**. This lets us keep the code easy to read.

Here is the C that determines the string of dashes and dots that you need to send to the **flashSequence** function:

```
letters[ch - 'a']
```

Once again, this looks a little strange. The function appears to be subtracting one character from another. This is actually a perfectly reasonable thing to do, because the function is actually subtracting the ASCII values.

Remember that you are storing the codes for the letters in an array. So the first element of the array contains a string of dashes and dots for the letter *A*, the second element includes the dots and dashes for *B*, and so on. So you need to find the right position in the array for the letter that you have just fetched from the buffer. The position for any lowercase letter will be the character code for the letter minus the character code for *a*. So, for example, *a* − *a* is actually 97 − 97 = 0. Similarly, *c* − *a* is actually 99 − 97 = 2. So, in the following statement, if ch **is** the letter c, then the bit inside the square brackets would evaluate to 2, and you would get element 2 from the array, which is -.-..

What this section has just described is concerned with lowercase letters. You also have to deal with uppercase letters and numbers. These are both handled in a similar manner.

The flashSequence Function

We have assumed a function called **flashSequence** and made use of it, but now you need to write it. We have planned for it to take a string containing a series of dashes and dots and to make the necessary flashes with the correct timings.

Thinking about the algorithm for doing this, you can break it into the following steps:

- For each element of the string of dashes and dots (such as .-.-):
  - Flash that dot or dash.

Using the concept of programming by intention, let's keep the function as simple as that.

The Morse codes are not the same length for all letters, so you need to loop around the string until you encounter the end marker, \0. You also need a counter variable called i that starts at 0 and is incremented as the processing looks at each dot and dash:

```
void flashSequence(char* sequence) {
    int i = 0;
    while (sequence[i] != '\0') {
        flashDotOrDash(sequence[i]);
        i++;
    }
    delay(dotDelay * 3);      // gap between letters
}
```

Again, you delegate the actual job of flashing an individual dot or dash to a new method called **flashDotOrDash**, which actually turns the LED on and off. Finally, when the program has flashed the dots and dashes, it needs to pause for three dots worth of delay. Note the helpful use of a comment.

The flashDotOrDash Function

The last function in your chain of functions is the one that actually does the work of turning the LED on and off. As its argument, the function has a single character that is either a dot (.) or a dash (–).

All the function needs to do is turn the LED on and delay for the duration of a dot if it's a dot and three times the duration of a dot if it's a dash, then turn the LED off again. Finally, it needs to delay for the period of a dot, to give the gap between flashes.

```
void flashDotOrDash(char dotOrDash) {
  digitalWrite(ledPin, HIGH);
  if (dotOrDash == '.') {
    delay(dotDelay);
  }
  else { // must be a -
    delay(dotDelay * 3);
  }
  digitalWrite(ledPin, LOW);
  delay(dotDelay); // gap between flashes
}
```

Putting It All Together

Putting all this together, the full listing is shown in sketch 04_05_morse_flasher. Upload it to your Arduino board and try it out. Remember that to use it, you need to open the Serial Monitor and type some text into the area at the top and click Send. You should then see that text being flashed as Morse code.

```
//sketch 04_05_morse_flasher
const int ledPin = 13;
const int dotDelay = 200;

char* letters[] = {
  ".-", "-...", "-.-.", "-..", ".", "..-.", "--.", "....", "..",    // A-I
  ".---", "-.-", ".-..", "--", "-.", "---", ".--.", "--.-", ".-.",  // J-R
  "...", "-", "..-", "...-", ".--", "-..-", "-.--", "--.."          // S-Z
};

char* numbers[] = {
  "-----", ".----", "..---", "...--", "....-", ".....", "-....", "--...",
  "---..", "----."};


void setup() {
  pinMode(ledPin, OUTPUT);
  Serial.begin(9600);
}

void loop() {
  char ch;
  if (Serial.available() > 0) {
    ch = Serial.read();
    if (ch >= 'a' && ch <= 'z') {
      flashSequence(letters[ch - 'a']);
    }
    else if (ch >= 'A' && ch <= 'Z') {
      flashSequence(letters[ch - 'A']);
    }
    else if (ch >= '0' && ch <= '9') {
      flashSequence(numbers[ch - '0']);
    }
    else if (ch == ' ') {
     delay(dotDelay * 4);        // gap between words
    }
  }
}

void flashSequence(char* sequence) {
    int i = 0;
    while (sequence[i] != NULL) {
        flashDotOrDash(sequence[i]);
        i++;
    }
    delay(dotDelay * 3);     // gap between letters
}
```

```
void flashDotOrDash(char dotOrDash) {
  digitalWrite(ledPin, HIGH);
  if (dotOrDash == '.') {
    delay(dotDelay);
  }
  else {
    // must be a dash
    delay(dotDelay * 3);
  }
  digitalWrite(ledPin, LOW);
  delay(dotDelay); // gap between flashes
}
```

This sketch includes a loop function that is called automatically and repeatedly calls a **flashSequence** function that you wrote, which itself repeatedly calls a **flashDotOrDash** function that you wrote, which calls **digitalWrite** and **delay** functions that are provided by Arduino!

This is how your sketches should look. Breaking things up into functions makes it much easier to get your code working and makes it easier when you return to it after a period of not using it.

The String Class

There are actually two ways to deal with strings of text in Arduino. There is the C character array that we have been using so far and there is also a String class (String with a capital S) that will look more like the stings you are used to if you have used almost any modern programming language. The problem with using the String class is that it generally uses more memory, which if you only have the 2k Bytes of an Arduino Uno can rapidly become a problem. However, when using a device with a lot more memory, the convenience that the String class offers when doing things like chopping parts out of a big string, or joining strings together, will simplify your code.

In Chapter 10, we will use the String class quite extensively.

Conclusion

In addition to looking at strings and arrays in this chapter, you have also built this more complex Morse translator that I hope will also reinforce the importance of building your code with functions.

In the next chapter, you learn about input and output, by which we mean input and output of analog and digital signals from the Arduino.

5
Input and Output

The Arduino is about physical computing, and that means attaching electronics to the Arduino board. So you need to understand how to use the various options for your connection pins.

Outputs can be digital, which just means switched between being at 0 V or at 5 V, or analog, which allows you to set the voltage to any voltage between 0 V and 5 V—although it's not quite as simple as that, as we shall see, and some boards use 3 V rather than 5 V.

Likewise, inputs can either be digital (for example, determining whether a button is pressed or not) or analog (such as from a light sensor).

In a book that is essentially about software rather than hardware, we are going to try and avoid being dragged into too much discussion of electronics. However, it will help you to understand what is happening in this chapter if you can find yourself a multimeter and a short length of solid core wire.

If you are interested in learning more about electronics, then you might like to look at my book *Hacking Electronics* (TAB/McGraw Hill, 2018).

Digital Outputs

In earlier chapters, you have made use of the light-emitting diode (LED) attached to digital pin 13 of the Arduino board. For example, in Chapter 4, you used it as a Morse code signaler. The Arduino board has a whole load of digital pins available.

Let's experiment with one of the other pins on the Arduino. You will use digital pin 4, and to see what is going on, you will fix some wire to your multimeter leads and attach them to your Arduino. Figure 5-1 shows the arrangement. If your multimeter has crocodile clips, strip the insulation off the ends of

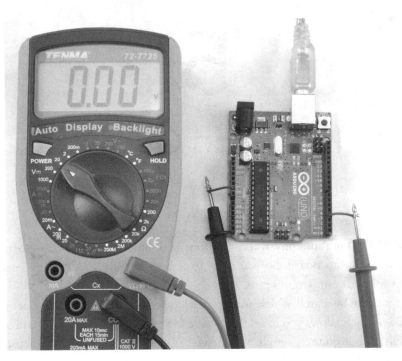

Figure 5-1 *Measuring outputs with a multimeter.*

some short lengths of solid core wire and attach the clip to one end, fitting the other end into the Arduino socket. If your multimeter does not have crocodile clips, then wrap one of the stripped wire ends around the probe as shown in Figure 5-1.

The multimeter needs to be set to its 0–20 V direct current (DC) range. The negative lead (black) should be connected to the ground (GND) pin and the positive to D3. The wire is just connected to the probe lead and poked into the socket headers on the Arduino board.

Load sketch 05_01_digital_out:

```
//05_01_digital_out
const int outPin = 3;

void setup() {
  pinMode(outPin, OUTPUT);
  Serial.begin(9600);
  Serial.println("Enter 1 or 0");
}
```

```
void loop() {
  if (Serial.available() > 0) {
    char ch = Serial.read();
    if (ch == '1') {
      digitalWrite(outPin, HIGH);
    }
    else if (ch == '0') {
      digitalWrite(outPin, LOW);
    }
  }
}
```

At the top of the sketch, you can see the command **pinMode**. You should use this command for every pin that you are using in a project so that Arduino can configure the electronics connected to that pin to be either an input or an output, as in the following example:

```
pinMode(outPin, OUTPUT);
```

As you might have guessed, **pinMode** is a built-in function. Its first argument is the pin number in question (an **int**), and the second argument is the mode, which must be either **INPUT, INPUT_PULLUP**, or **OUTPUT**. Note that the mode name must be all uppercase.

This **loop** waits for a command of either **1** or **0** to come from the Serial Monitor on your computer. It it's a **1**, then pin 3 will be turned on; otherwise, it will be turned off.

Upload the sketch to your Arduino and then open the Serial Monitor (shown in Figure 5-2).

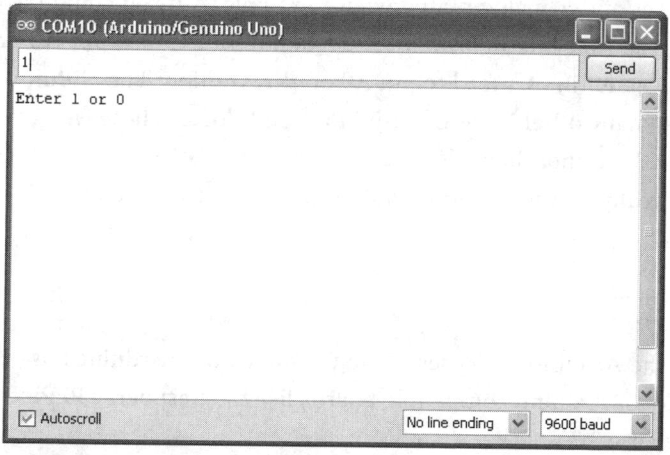

Figure 5-2 *The Serial Monitor.*

Figure 5-3 *Setting the output to High.*

So, with the multimeter turned on and plugged into the Arduino, you should be able to see its reading switch between 0 V and about 5 V as you send commands to the board from the Serial Monitor by either pressing 1 and then ENTER or pressing 0 and then ENTER. Figure 5-3 shows the multimeter reading after a 1 has been sent from the Serial Monitor.

If there are not enough pins labeled "D" for your project, you can actually use the pins labeled "A" (for analog) as digital outputs too. To do this, just use the letter A in front of the analog pin name, for example, A0. You could try this out by modifying the first line in sketch 05_01_digital_out to use pin A0 and moving your positive multimeter lead to pin A0 on the Arduino. The Arduino Uno's pins are 5 V, but many other types of Arduino use 3 V logic. These will give an output voltage of 3.3 V rather than 5 V.

That is really all there is to digital outputs, so let's move on swiftly to digital inputs.

5V or 3.3V?

Although the Arduino Uno uses 5 V logic, some other Arduinos use 3.3 V rather than 5 V. Most Arduino compatible boards like the Raspberry Pi Pico and ESP32 boards also use 3.3 V rather than 5 V.

Digital Inputs

The most common use of digital inputs is to detect when a switch has been closed. A digital input can either be on or off. If the voltage at the input is less than 2.5 V (halfway to 5 V), it will be 0 (off), and if it is above 2.5 V, it will be 1 (on).

Disconnect your multimeter and upload sketch 05_02_digital_input onto your Arduino board:

```
//05_02_digital_input
const int inputPin = 5;

void setup() {
  pinMode(inputPin, INPUT);
  Serial.begin(9600);
}

void loop() {
  int reading = digitalRead(inputPin);
  Serial.println(reading);
  delay(1000);
}
```

As with using an output, you need to tell the Arduino in the **setup** function that you are going to use a pin as an input. You get the value of a digital input using the **digitalRead** function. This returns 0 or 1.

Pull-Up Resistors

The sketch reads the input pin and writes its value to the Serial Monitor once per second. So upload the sketch and open the Serial Monitor. You should see a value appear once per second. Push one end of your bit of wire into the socket for D5 and pinch the end of the wire between your fingers, as shown in Figure 5-4.

Continue pinching for a few seconds and watch the text appear on the Serial Monitor. You should see a mixture of ones and zeros appear in the Serial Monitor. The reason for this is that the inputs to the Arduino board are very sensitive. You are acting as an antenna, picking up electrical interference.

Take the end of the wire that you were holding and push it into the socket for +5 V as shown in Figure 5-5. The stream of text in the Serial Monitor should change to ones.

Figure 5-4   *A digital input with a human antenna.*

Figure 5-5   *Pin 5 connected to +5V.*

Figure 5-6   *Connecting a switch to an Arduino board.*

Now take the end that was in +5 V and put it into one of the GND connections on the Arduino. As you would expect, the Serial Monitor should now display zeros.

A typical use for an input pin is to connect a switch to it. Figure 5-6 shows how you might connect your switch.

The problem with this is that if the switch is not closed, then the input pin is not connected to anything. It is said to be floating, and could easily give you a false reading. You need your input to be more predictable, and the way to do this is with what is called a pull-up resistor. You will see later how you can enable the Arduino's built-in series resistors and avoid having to use separate resistors. Figure 5-7 shows the standard use of a pull-up resistor. It has the effect that if the switch is open, then the resistor pulls up the floating input to 5 V. When you press the switch and close the contact, the switch overrides the effect of the resistor, forcing the input to 0 V. One side-effect of this is, while the switch is closed, 5 V will be across the resistor, causing a current to flow. So, the value of the resistor is selected to be low enough to make it immune from any electrical

Figure 5-7 *Switch with a pull-up resistor.*

interference, but at the same time high enough to prevent excessive current drain when the switch is closed.

Internal Pull-Up Resistors

Fortunately, the Arduino board has software-configurable pull-up resistors built into the digital pins. By default, they are turned off. So all you need to do to enable the pull-up resistor on pin 5 for sketch 05_02_digital_input is to change the pin mode from INPUT to INPUT_PULLUP.

Sketch 05_03_digital_input_pullup is the modified version. Upload it to your Arduino board and test it by acting like an antenna again. You should find that this time the input stays at 1 in the Serial Monitor.

```
//05_03_digital_input_pullup
const int inputPin = 5;

void setup() {
  pinMode(inputPin, INPUT_PULLUP);
  Serial.begin(9600);
}

void loop() {
  int reading = digitalRead(inputPin);
  Serial.println(reading);
  delay(1000);
}
```

Debouncing

When you press a push button, you would expect that you would just get a single change from 1 (with a pull-up resistor) to 0 as the button is depressed. Figure 5-8

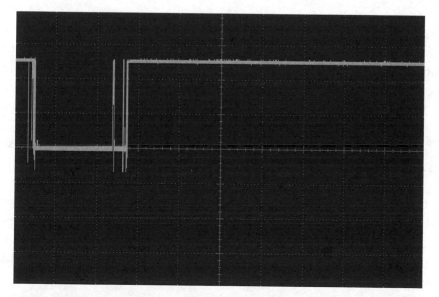

Figure 5-8 *Oscilloscope trace of a button press.*

shows what can happen when you press a button. The metal contacts in the button bounce. So a single button press becomes a series of presses that eventually stabilize.

All this happens very quickly; the total time span of the button press on the oscilloscope trace is only 200 milliseconds. This is a very "ropey" old switch. A new tactile, click-type button may not even bounce at all.

Sometimes bouncing does not matter. For instance, sketch 05_04_switch_led will light the LED while the button is pressed. In reality, you would not use an Arduino to do this; we are firmly in the realms of theory rather than practice here.

```
//05_04_switch_led
const int inputPin = 5;
const int ledPin = 13;

void setup() {
  pinMode(ledPin, OUTPUT);
  pinMode(inputPin, INPUT_PULLUP);
}

void loop() {
  int switchOpen = digitalRead(inputPin);
  digitalWrite(ledPin, ! switchOpen);
}
```

Looking at the **loop** function of sketch 05_04_switch_led, the function reads the digital input and assigns its value to a variable **switchOpen**. This is a 0 if the button is pressed and a 1 if it isn't (remember that the pin is pulled up to 1 when the button is not pressed).

When you program **digitalWrite** to turn the LED on or off, you need to reverse this value. You do this using the **!** or **not** operator.

If you upload this sketch and connect your wire between D5 and GND (see Figure 5-9), you should see the LED light. Bouncing may be going on here, but it is probably too fast for you to see and does not matter.

One situation where key bouncing would matter is if you were making your switch toggle the LED on and off. That is, if you press the button, the LED comes on and stays on, and when you press the button again, it turns off. If you had a button that bounced, then whether the LED was on or off would just depend on whether you had an odd or even number of bounces.

Figure 5-9 *Using a wire as a switch.*

Sketch 05_05_toggle just toggles the LED without any attempt at "debouncing." Try it out using your wire as a switch between pin D5 and GND (or use a real switch if you have one):

```
//05_05_toggle
const int inputPin = 5;
const int ledPin = 13;
int ledValue = LOW;

void setup() {
  pinMode(inputPin, INPUT_PULLUP);
  pinMode(ledPin, OUTPUT);
}

void loop() {
  if (digitalRead(inputPin) == LOW) {
    ledValue = ! ledValue;
    digitalWrite(ledPin, ledValue);
  }
}
```

You will probably find that sometimes the LED toggles, but other times it appears not to toggle. This is bouncing in action!

A simple way to tackle this problem is simply to add a delay after you detect the first button press, as shown in sketch 05_06_bounce_delay:

```
//05_06_bounce_delay
const int inputPin = 5;
const int ledPin = 13;
int ledValue = LOW;

void setup() {
  pinMode(inputPin, INPUT_PULLUP);
  pinMode(ledPin, OUTPUT);
}

void loop() {
  if (digitalRead(inputPin) == LOW) {
    ledValue = ! ledValue;
    digitalWrite(ledPin, ledValue);
    delay(500);
  }
}
```

By putting a delay here, nothing else can happen for 500 milliseconds, by which time any bouncing will have subsided. You should find that this makes the toggling much more reliable. An interesting side-effect is that if you hold the button down, the LED just keeps on flashing.

If that is all there is to the sketch, then this delay is not a problem. However, if you do more in the **loop**, then using a delay can be a problem; for example, the program would be unable to detect the press of any other button during that 500 milliseconds.

So, this approach is sometimes not good enough and you will need to be a bit more sophisticated. You can write your own advanced debouncing code by hand, but doing so gets complicated and fortunately some fine folks have done all the work for you.

To make use of their work, you must add a library to your Arduino application. To do this, open the Library Manager (Figure 5-10) from Sketch | Include Library | Manage Libraries... menu. Then type Bounce2 into the search field. This should bring the library Bounce2 to near the top of the results. Select it and then click Install.

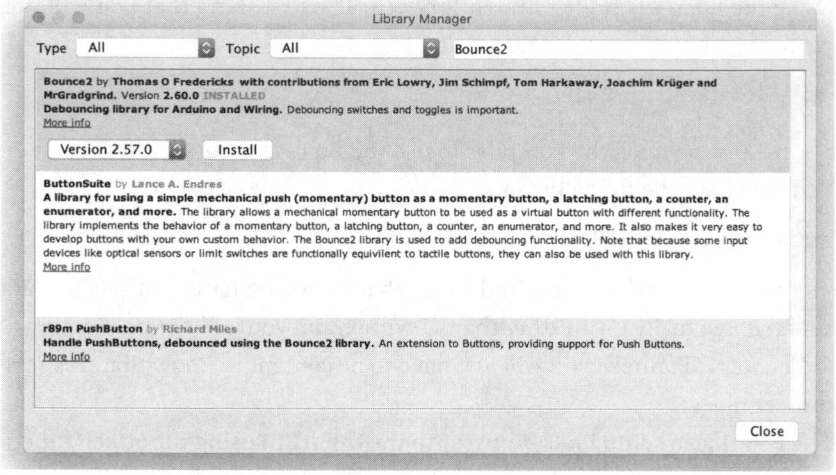

Figure 5-10 *Installing the Bounce library.*

Sketch 05_07_bounce_library shows how you can use the Bounce library. Upload it to your board and see how reliable the LED toggling has become.

```
//05_07_bounce_library
#include <Bounce2.h>

const int inputPin = 5;
const int ledPin = 13;

int ledValue = LOW;
Bounce bouncer = Bounce();

void setup() {
  pinMode(inputPin, INPUT_PULLUP);
  pinMode(ledPin, OUTPUT);
  bouncer.attach(inputPin);
}

void loop() {
  if (bouncer.update() && bouncer.read() == LOW) {
   ledValue = ! ledValue;
   digitalWrite(ledPin, ledValue);
  }
}
```

Using the library is pretty straightforward. The first thing that you will notice is this line:

```
#include <Bounce2.h>
```

This is necessary to tell the compiler to use the Bounce library.

You then have the following line:

```
Bounce bouncer = Bounce();
```

Do not worry about the syntax and somewhat sing-song nature of this line at the moment; it is actually C++ rather than C syntax, and you will not be meeting C++ until Chapter 7. For now, you will just have to be content to know that this sets up a **bouncer** object.

The new line in setup links **bouncer** to the **inputPin** using the attach function. From now on, you use that **bouncer** object to find out what the switch is doing rather than reading the digital input directly. It has put a kind of debouncing wrapper around your input pin. So, deciding whether a button has been pressed is wrapped up in this line:

```
if (bouncer.update() && bouncer.read() == LOW)
```

The function **update** returns true if something has changed with the **bouncer** object and the second part of the condition checks whether the button went **LOW**.

Analog Outputs

A few of the digital pins of an Arduino Uno—namely digital pins 3, 5, 6, 9, 10, and 11—can provide variable output other than just 5 V or nothing. These are the pins on the board with a ~ or "PWM" next to them. PWM stands for Pulse Width Modulation, which refers to the means of controlling the amount of power at the output. It does so by rapidly turning the output on and off. Other types of board may have PWM available on different pins and sometimes on all pins.

On an Arduino Uno, the pulses are always delivered at the same rate (roughly 500 per second on all the pins except pins 5 and 6, which provide 980 pulses per second), but the length of the pulses is varied. If you were to use PWM to control the brightness of an LED, then if the pulse were long, your LED would be on all the time. If, however, the pulses are short, then the LED is actually lit only for a

Figure 5-11 *Measuring the analog output.*

small portion of the time. This happens too fast for the observer even to tell that the LED is flickering, and it just appears that the LED is lighter or dimmer.

Before you try using an LED, you can test this out with your multimeter. Set the multimeter up to measure the voltage between GND and pin D3 (see Figure 5-11).

Now upload sketch 05_08_analog_output to your board and open the Serial Monitor (see Figure 5-12). Enter **3** and press ENTER. You should see your volt meter register about 3 V. You can then try any other voltage between 0 and 5.

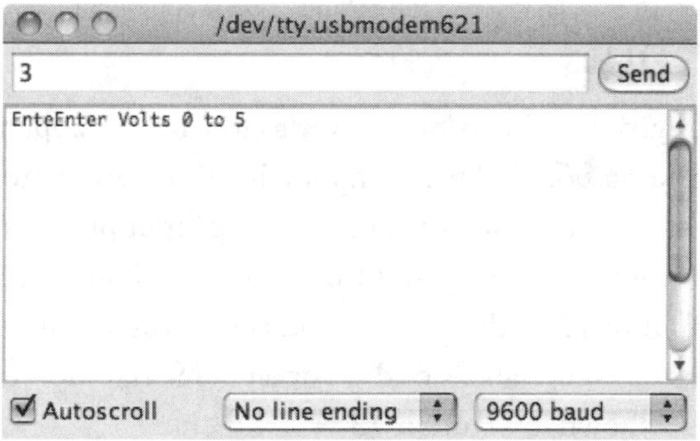

Figure 5-12 *Setting the voltage at an analog output.*

```
//05_08_analog_output
const int outputPin = 3;

void setup() {
  pinMode(outputPin, OUTPUT);
  Serial.begin(9600);
  Serial.println("Enter Volts 0 to 5");
}

void loop() {
  if (Serial.available() > 0) {
    float volts = Serial.parseFloat();
    int pwmValue = volts * 255.0 / 5.0;
    analogWrite(outputPin, pwmValue);
  }
}
```

The program determines the value of PWM output between 0 and 255 by multiplying the desired voltage (0 to 5) by 255/5. (Readers may wish to refer to Wikipedia for a fuller description of PWM.)

You can set the value of the output by using the function **analogWrite**, which requires an output value between 0 and 255, where 0 is off and 255 is full power. This is actually a great way to control the brightness of an LED. If you were to try to control the brightness by varying the voltage across the LED, you would find that nothing would happen until you got to about 2 V; then the LED would very quickly get quite bright. By controlling the brightness using PWM and varying the average amount of time that the LED is on, you achieve much more linear control of the brightness.

Analog Input

Digital inputs just give you an on/off answer as to what is happening at a particular pin on the Arduino board. Analog inputs, however, give you a value between 0 and 1023 depending on the voltage at the analog input pin.

The program reads the analog input using the **analogRead** function. Sketch 05_09_analog_input displays the reading and actual voltage at the analog pin A0 in the Serial Monitor every half second, so open the Serial Monitor and watch the readings appear, as shown in Figure 5-13.

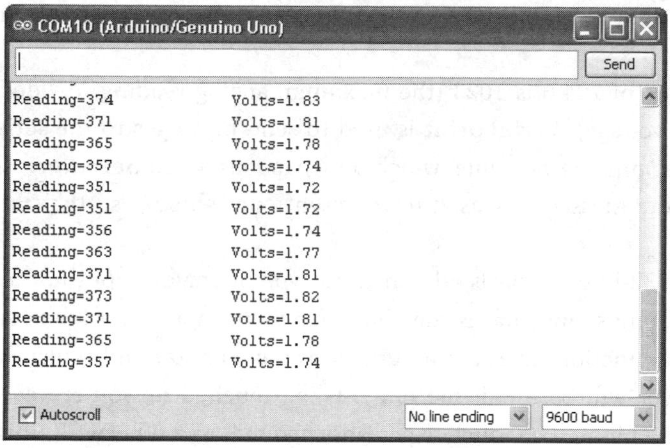

```
COM10 (Arduino/Genuino Uno)                              [_][□][X]
|                                                        [ Send ]
Reading=374              Volts=1.83
Reading=371              Volts=1.81
Reading=365              Volts=1.78
Reading=357              Volts=1.74
Reading=351              Volts=1.72
Reading=351              Volts=1.72
Reading=356              Volts=1.74
Reading=363              Volts=1.77
Reading=371              Volts=1.81
Reading=373              Volts=1.82
Reading=371              Volts=1.81
Reading=365              Volts=1.78
Reading=357              Volts=1.74
[✓] Autoscroll          [ No line ending ▾ ] [ 9600 baud ▾ ]
```

Figure 5-13 *Measuring voltage with an Arduino Uno.*

```
//05_09_analog_input
const int analogPin = 0;

void setup() {
  Serial.begin(9600);
}

void loop() {
  int reading = analogRead(analogPin);
  float voltage = reading / 204.6;
  Serial.print("Reading=");
  Serial.print(reading);
  Serial.print("\t\tVolts=");
  Serial.println(voltage);
  delay(500);
}
```

When you run this sketch, you will notice that the readings change quite a bit. As with the digital inputs, this is because the input is floating.

Take one end of the wire and put it into a GND socket so that A0 is connected to GND. Your readings should now stay at 0. Move the end of the lead that was in GND and put it into 5 V and you should get a reading of around 1023, which is the maximum reading. So, if you were to connect A0 to the 3.3 V socket on the Arduino board, the Arduino voltmeter should tell you that you have about 3.3 V.

The value of 204.6 is 1023 (the maximum analog reading) divided by 5 (the maximum voltage). **Serial.print** is used to send messages to the serial monitor without beginning a new line, which only happens when **Serial.println** is used. The \t in the messages is used to represent one tab stop so that the numbers line up.

The ESP32-based boards add an extra type of analog input not available on official Arduinos, and that is the ability for a pin to act as a touch sensor. The **touchRead** function takes a pin number as a parameter and returns an integer. This number will be low if the pin is being touched or you are even close to touching the pin or a conductive pad attached to it and otherwise a higher value. Depending on what's connected to your pin, you can choose a threshold to decide if the touch button has been pressed or not.

If you are using an Arduino with 3 V logic, then instead of dividing by 204.6, to get the voltage, you would divide by 310 (1023/3.3).

Conclusion

This concludes our chapter on the basics of getting signals into and out of the Arduino. In the next chapter, we will look at some more advanced Arduino features.

6

Boards

The Arduino Uno is probably the best board to use when learning Arduino. It is the closest thing there is to a standard Arduino. However, as you start to make projects for real, embedding an Arduino Uno into each project gets expensive. What's more, some projects will have special requirements, such as the need to be very compact or need to use WiFi or Bluetooth. For such projects, a host of other Arduino compatible boards are available. They can still be programmed from the Arduino IDE, so all that you have learnt about programming still applies.

There are many more boards available than those listed here, but these make a fairly representative sample of the kinds of boards available.

Arduino Nano

The Arduino Nano (see Figure 6-1) is essentially an Arduino Uno shrunk down to be as compact as possible. It uses the same microcontroller as the Uno (ATmega328) but the header sockets of the Uno are replaced by header pins.

Because the Arduino Nano is so similar to the Uno, it is a great alternative if you need your project to be small.

When programming an Arduino Nano, you need to select that as the board type from the Tools menu of the Arduino IDE (Figure 6-2).

The official Arduino boards are expensive. Yes, they are of high quality and made in Italy, and nicely packaged, but as microcontroller boards go, they are expensive. Arduino boards have an open source design. That is, the design files used to make

Figure 6-1 *The Arduino Nano.*

Tools	Help		
Auto Format	⌘T	Boards Manager...	
Archive Sketch			
Fix Encoding & Reload		Arduino AVR Boards	
Manage Libraries...	⇧⌘I	Arduino Yún	
Serial Monitor	⇧⌘M	Arduino/Genuino Uno	
Serial Plotter	⇧⌘L	Arduino Duemilanove or Diecim	
		✓ Arduino Nano	
WiFi101 / WiFiNINA Firmware Updater		Arduino/Genuino Mega or Meg:	
		Arduino Mega ADK	
Board: "Arduino Nano"	▶	Arduino Leonardo	
Processor: "ATmega328P"	▶	Arduino Leonardo ETH	
Port	▶	Arduino/Genuino Micro	
Get Board Info		Arduino Esplora	

Figure 6-2 *Selecting the board type in the Arduino IDE.*

them are made public. This has enabled Chinese manufacturers to make versions of the Nano and other Arduino boards that cost a small fraction of the official boards. These copies are inferior in several ways. The boards are often literally a bit rough around the edges, and often seconds of the microcontroller chip are used and the USB interface chip substituted by a different (and cheaper) USB interface chip. This last point means that often, when using Windows, drivers for these chips have to be installed before the board will be recognized by the Arduino IDE.

Arduino Pro Mini

The Arduino Uno and Pico both have USB interface chips that serve the dual purposes of providing a means of programming the Arduino and a way for the Arduino to pass data over USB to a computer. Many Arduino projects do not require this second feature, and so, for those projects, the USB interface is only really needed while programming the Arduino.

The Arduino Pro Mini (Figure 6-3) is like an Arduino Nano, but to keep the cost down the USB interface is replaced by a serial interface and therefore doesn't require an expensive USB interface chip. This makes the Pro Mini cheaper than the Arduino Nano but it does mean that you need a separate USB to serial adaptor for it.

The Pro Mini is available in two versions, 5 V and 3 V. The 5 V version is closer to the Arduino Uno, running at 16 MHz, whereas the 3 V version has an 8 MHz clock speed. This is a trade-off between operating voltage and speed imposed by the microcontroller of the boards. To operate reliably at 16 MHz the microcontroller requires a 5 V supply.

As with the Arduino Nano, there is no end of cheap copies of the Pro Mini.

Figure 6-3   *An Arduino Pro Mini and USB to serial adaptor.*

Breadboard

Most of the boards that we will cover in this chapter have a similar pin arrangement to an Arduino Pro Mini or Nano. That is, they are tall and thin in aspect ratio with a row of pins down each side. These pins are 0.1 inches apart and are designed that way, largely so that they can be used with solderless breadboard (often just called breadboard).

Figure 6-4 shows an Arduino Nano on solderless breadboard with a light-emitting diode (LED) and resistor also on the breadboard.

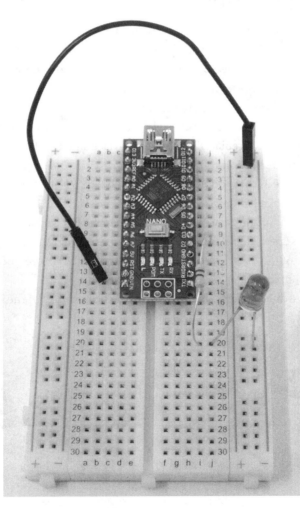

Figure 6-4 *An Arduino Nano on breadboard.*

Solderless breadboard is a great way to prototype your projects, because it allows you to easily connect things to your board without the need for any soldering. While you can connect breadboard to an Arduino Uno using jumper wires, it is easy for them to become separated as the Uno and breadboard are not physically connected.

Behind each row of holes on the breadboard lies a clip that grips any wires or component legs pushed through the breadboard.

The Boards Manager

A fresh install of the Arduino IDE will reveal a long list of boards when you go to the Tools and then Board menu options (Figure 6-5). These are just the official Arduino boards, and you can see the Arduino Uno, Nano, and Pro Mini there. So when using one of these boards, you have to select it from this list before you can upload a program to it.

At the top of the boards list is the Boards Manager option. This allows you to tell the Arduino IDE about all sorts of other boards. This allows anyone (with

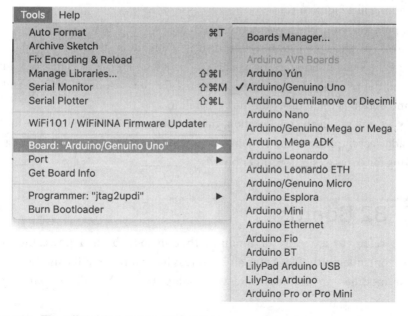

Figure 6-5   *The official Arduino board options.*

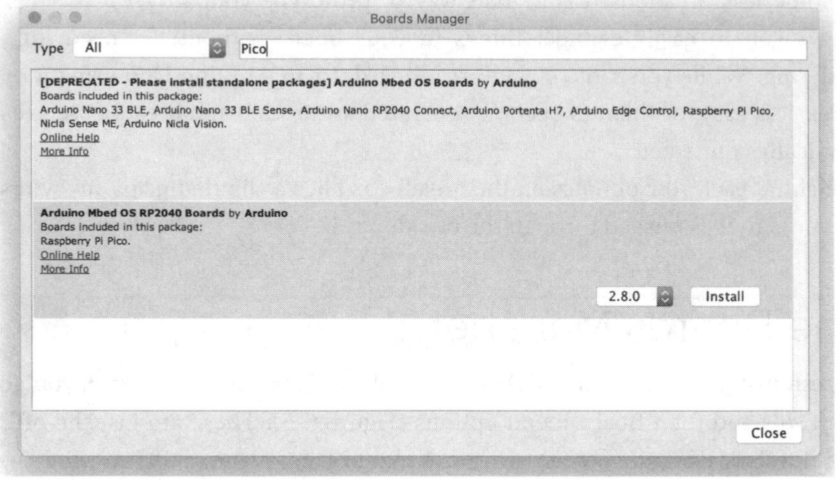

Figure 6-6 *The board manager.*

quite a lot of work) to add their own boards to be programmed by the Arduino IDE. This open feature of the Arduino IDE is one of the main reasons for its popularity, as the Arduino community has made boards using all types of microcontroller accessible. When you open the Boards Manager, you will see something like Figure 6-6.

Typing the name of a board in the search area will find that board if it is known and give you the option to install it, or if it is already installed to uninstall it of change the version. Support for a board is termed a *core*. So, in Figure 6-5, I have entered the text Pico hoping to find support for the Raspberry Pi Pico and the second option is there ready for me to install it.

As we will see in the next section, this list of boards is not exhaustive and we can add search paths to the Arduino IDE, telling it other places where it can go looking for cores to support other boards.

ESP32 Boards

ESP32 refers to a family of modules that include both a powerful 32-bit microcontroller with GPIO pins and wireless hardware for WiFi and Bluetooth. These modules are low cost and when placed on a board, provide a great

Figure 6-7 *A typical ESP32 board.*

alternative to an Arduino Nano, especially when you need to use WiFi or Bluetooth in your project. There are lots of manufacturers of boards using the ESP32 modules. Typical of these are ESP32 Wroom and the Wemos LOLIN32 Lite shown in Figure 6-7. All of these boards operate at 3 V and NOT the 5 V of an Arduino Uno.

This particular board is not included in the list of boards known to the Boards Manager. To allow the Boards Manager to find it, we have to add a web address to the Arduino IDE's configuration. So, open the Arduino IDE Preferences panel from the menu (Figure 6-8) and paste the following URL into the Additional Boards Manager URLs field.

https://dl.espressif.com/dl/package_esp32_index.json

Now, open the Boards Manager and type ESP32 into the search field (Figure 6-9).

After the ESP32 core has finished installing (which may take a while), you should find a load more options in your Boards menu.

Transferring a program to Arduino Uno is really quick, although it gets slower the bigger the sketch. In contrast, uploading a program for an ESP32-based board will take a lot longer. If uploading fails for one off these boards, try reducing "Upload speed" in the Tools menu.

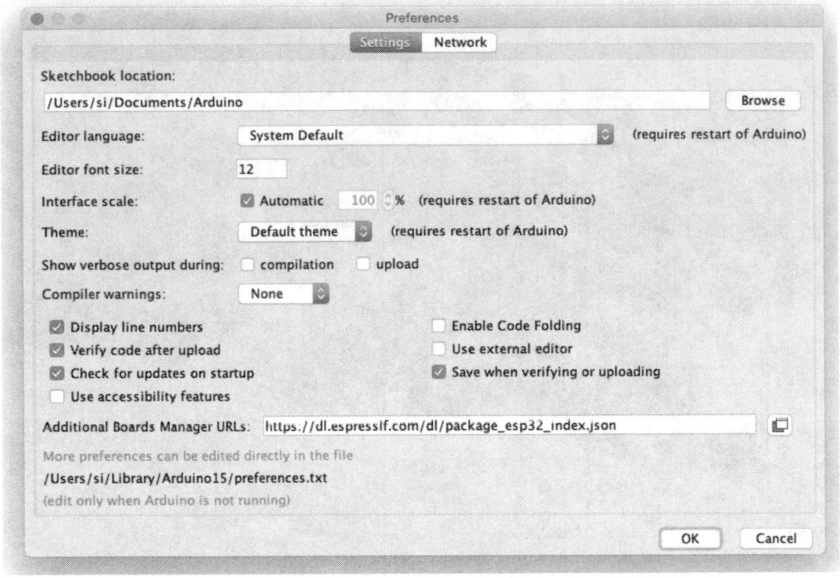

Figure 6-8   *Adding a URL to preferences.*

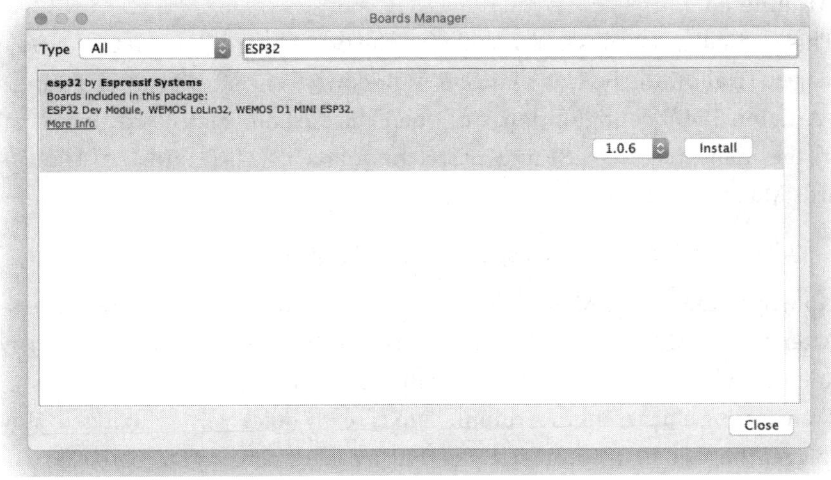

Figure 6-9   *Installing support for ESP32 boards.*

A predecessor to the ESP32 is the ESP8266. These are similar modules, but considerably less powerful. But for the small amount of extra money, you may as well use the newer ESP32. Some manufacturers even include displays and other peripherals such as long-range radio and OLED displays on their ESP32 boards.

You will meet an ESP32 board again in Chapter 10 when we use one with WiFi.

Raspberry Pi Pico

The Raspberry Pi Pico (Figure 6-10) is not to be confused with a regular Raspberry Pi which is a single board computer. The Raspberry Pi Pico is a microcontroller board in a breadboard friendly format. Although the manufacturers of the Pico intended it to be primarily programmed using the Python programming language, the Arduino organization has taken the microcontroller chip (RP2040) and put it into an official Arduino board. So, you can use the official Arduino board or a much cheaper Raspberry Pi Pico in your projects. This means that the Arduino IDE also supports the Pico through an official core. However, at the time of

Figure 6-10 *The Raspberry Pi Pico.*

writing, an unofficial core developed by Earle Philhower supports more Pico-type boards and works very well.

The Pico does not have any extra hardware features like an ESP32-based device, but it is remarkably low-cost and has a very impressive processor of similar power to the ESP32. So, if you want the power of an ESP32 but don't need WiFi and Bluetooth, then the Pico makes a good Arduino board.

Installing the unofficial support for the Pico board into the Arduino IDE is a similar process to installing support for ESP32. You first have to open Preferences and then add the following URL into the Additional Boards Manager URLs field.

```
https://github.com/earlephilhower/arduino-pico/releases/
download/global/package_rp2040_index.json
```

You can copy this URL from the project page at https://github.com/earlephilhower/arduino-pico. If you already have a URL (such as the ESP32 URL) in the preferences field, then click on the icon just after the field to open it in a multi-line edit window, where you can add the URL on a line of its own.

Open the Boards Manager and search for Pico, this will bring up the official Arduino Pico core as well as the core you want, from Earle Philhower.

For whatever reason, the Pico's designers decided to write the pin numbers on the underside of the board. This is not very convenient if you are using breadboard. One way to make pin identification easier is to use the MonkMakes Breadboard for Pico (https://monkmakes.com/pico_bb). This is normal breadboard, but with the Pico pin names written on the breadboard.

BBC micro:bit

The BBC micro:bit (Figure 6-11) is a very interesting board, that is extremely popular as an educational tool to teach programming and electronics. What distinguishes the board from the boards that we have seen so far is that the micro:bit comes with a number of peripherals on the board:

- A 5 × 5 LED display
- A small speaker (from version 2 of the micro:bit)
- A microphone (from version 2 of the micro:bit)
- An accelerometer (to detect movement)
- A magnetometer

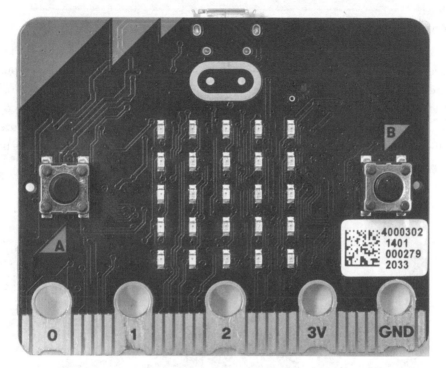

Figure 6-11 *The BBC micro:bit.*

By a clever trick, it can use the LEDs to measure the light level and can also report the temperature of its processor chip.

Although the micro:bit is most often programmed using the Makecode blocks-based programming environment, it's another board like the Pico, that you can program from the Arduino IDE. To do so, you need to add the URL below to the Additional Boards Manager URLs field in Preferences before searching for micro:bit and installing the core called *Nordic Semiconductor nRF5 Boards by Sandeep Mistry*.

https://sandeepmistry.github.io/arduino-nRF5/package_nRF5_boards_index .json

To make full use of all the micro:bit's peripherals, you will have to download other Arduino libraries. You can find out more about this here: https://learn .adafruit.com/use-micro-bit-with-arduino

Adafruit Feather System

Adafruit have formalized a style of board that the call Feathers (for example, Figure 6-12). These boards are all the same size and have the same basic pinout, but are available in a huge range of microcontroller variants, including the RP2040 (of the Raspberry Pi Pico), ESP32, and many others. In many ways this rivals the ecosystem of the original Arduino, especially as it includes Feather Wings, that are the same idea as Arduino Shields, adding extra hardware features to the boards in a plug-in manner.

You can find out more about the Feather system here: https://www.adafruit.com/category/943

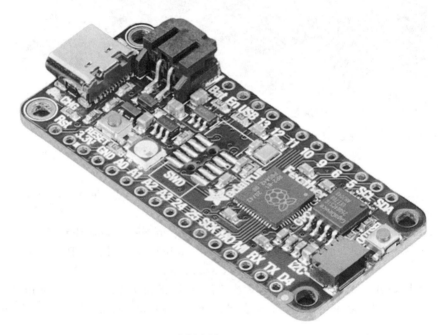

Figure 6-12   *The Adafruit Feather RP2040.*

Conclusion

There are so many boards, that it is somewhat unfair to recommend just a few boards. However Table 6-1 should point you in the right general direction, at least with the boards I've described here.

Board	Features
Arduino Uno	Easy to get started with. The standard Arduino.
Arduino Nano	A compact version of the Uno, suitable for breadboard use.
Arduino Pro Mini	A lower cost version of the Nano, without a USB interface. Great for embedding in final projects.
ESP32 board	WiFi, Bluetooth, fast processor, breadboard format at a cost low enough to embed in projects. Slow to program.
Raspberry Pi Pico	Low cost and powerful processor.
BBC micro:bit	Lots of built-in peripherals.

Table 6-1 *Summary of Boards*

7

Advanced Arduino

In this chapter, we will look at some of the more advanced features of Arduino that we have not yet come across and take a more in-depth look at the Arduino's standard library of functions and data types.

You have already met a fair few of the built-in functions, such as **pinMode**, **digitalWrite**, and **analogWrite**. But actually, there are many more. There are functions that you can use for doing math, making random numbers, manipulating bits, detecting pulses on an input pin, and using something called interrupts.

The Arduino language is based on an earlier library called Wiring and it complements another library called Processing. The Processing library is very similar to Wiring, but it is based on the Java language rather than C and is used on your computer to link to your Arduino over USB. In fact, the Arduino IDE application that you run on your computer is based on Processing. If you find yourself wanting to write some fancy interface on your computer to talk to an Arduino, then take a look at Processing (www.processing.org).

Random Numbers

Despite the experience of anyone using a PC, computers are in actual fact very predictable. Occasionally it is useful to be able to deliberately make your Arduino unpredictable. For example, you might want to make a robot take a "random" path around a room, heading for a random amount of time in one direction, turning a random number of degrees, and then setting off again. Or, you might be contemplating making an Arduino-based die that gives you a random number between one and six.

The Arduino standard library provides you with a feature to do just this. It is the function called **random**. **random** returns an **int** and it can take either one argument or two. If it just takes one argument, then it will return a random number between zero and the argument minus one.

The two-argument version produces a random number between the first argument (inclusive) and the second argument minus one. Thus **random(1, 10)** produces a random number between one and nine.

Sketch 07_01_random pumps out numbers between one and six to the Serial Monitor.

```
//sketch 07_01_random

void setup() {
  Serial.begin(9600);
}

void loop() {
  int number = random(1, 7);
  Serial.println(number);
  delay(500);
}
```

If you upload this sketch to your Arduino and open the Serial Monitor, you will see something like Figure 7-1.

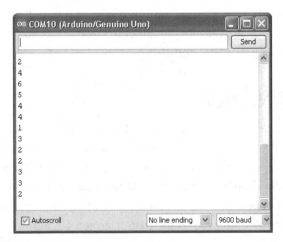

Figure 7-1 *Random numbers.*

If you run this a few times you will probably be surprised to see that every time you restart your Arduino you get the same series of "random" numbers.

The output is not really random; the numbers are called *pseudo-random* numbers because they have a random distribution. That is, if you ran this sketch and collected a million numbers, you would get pretty much the same number of ones, twos, threes, and so on. The numbers are not random in the sense of being unpredictable. In fact, it is so against the workings of a microcontroller to be random that it just plain can't do it without some intervention from the real world.

You can provide this intervention to make your sequence of numbers less predictable by *seeding* the random number generator. This basically just gives it a starting point for the sequence. But, if you think about it, you cannot just use **random** to seed the random number generator. A commonly used trick is to use the fact that (as discussed in the last chapter) an analog input will float. So you can use the value read from an analog input to seed the random number generator.

The function that does this is called **randomSeed**. Sketch 07_02_random_seed shows how you can add a bit more randomness to your random number generator.

```
//sketch 07_02_random_seed

void setup() {
  Serial.begin(9600);
  randomSeed(analogRead(0));
}

void loop() {
  int number = random(1, 7);
  Serial.println(number);
  delay(500);
}
```

Try pressing the Reset button on your Arduino a few times. You should now see that your random sequence is different every time.

This type of random number generation could not be used for any kind of lottery. For much better random number generation, you would need hardware random number generation, which is sometimes based on random occurrences, such as cosmic ray events.

Math Functions

On rare occasions, you will need to do a lot of math on an Arduino, over and above the odd bit of arithmetic. But, should you need to, there is a big library of math functions available to you. The most useful of these functions are summarized in the following table.

Function	Description	Example
abs	Returns the unsigned value of its argument.	**abs(12) returns 12** **abs(-12) returns 12**
constrain	Constrains a number to stop it from exceeding a range. The first argument is the number to constrain, the second is the start of the range, and the third is the end of the allowed range of numbers.	**constrain(8, 1, 10) returns 8** **constrain(11, 1, 10) returns 10** **constrain(0, 1, 10) returns 1**
map	Maps a number in one range into another range. The first argument is the number to map, the second and third are the "from" range (or source range), and the last two are the "to" range (or destination range). The function is useful for remapping analog input values.	**map(x, 0, 1023, 0, 5000)**
max	Returns the larger of its two arguments.	**max(10, 11) returns 11**
min	Returns the smaller of its two arguments.	**min(10, 11) returns 10**
pow	Returns the first argument raised to the power of the second argument.	**pow(2, 5) returns 32**
sqrt	Returns the square root of a number.	**sqrt(16) returns 4**
sin, **cos**, **tan**	Perform trigonometric functions. They are not often used.	
log	Calculates the temperature from a logarithmic thermistor (for example).	

Bit Manipulation

A bit is a single digit of binary information, that is, either 0 or 1. The word *bit* is a contraction of *binary digit*. Most of the time, you use **int** variables that actually comprise 16 bits. This is a bit wasteful if you only need to store a simple true/false value (1 or 0). Actually, unless you are running short of memory, being wasteful is less of a problem than creating difficult-to-understand code, but sometimes it is useful to be able to pack your data tightly.

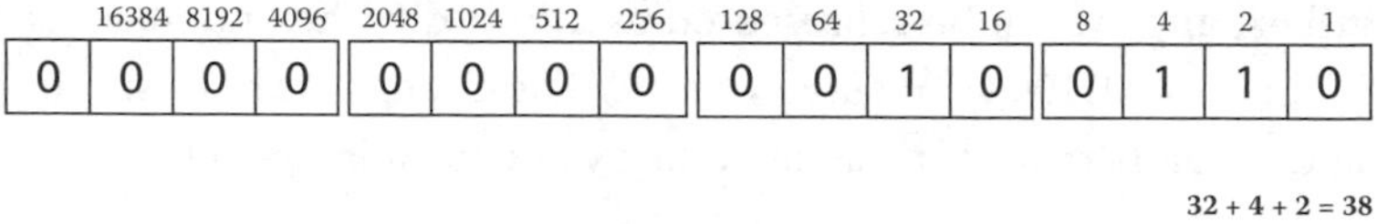

16384	8192	4096	2048	1024	512	256	128	64	32	16	8	4	2	1
0	0	0	0	0	0	0	0	0	0	1	0	0	1	1	0

$$32 + 4 + 2 = 38$$

Figure 7-2 *An **int**.*

Each bit in the **int** can be thought of as having a decimal value, and you can find the decimal value of the **int** by adding up the values of all the bits that are a 1. So in Figure 7-2, the decimal value of the **int** would be 38. Actually, it gets more complicated to deal with negative numbers, but that only happens when the leftmost bit becomes a 1.

When you are thinking about individual bits, decimal values do not really work very well. It is very difficult to visualize which bits are set in a decimal number such as 123. For that reason, programmers often use something called *hexadecimal*, or, more commonly, just *hex*. Hex is number base 16. So instead of having digits 0 to 9, you have six extra digits, A to F. This means that each hex digit represents four bits. The following table shows the relationship among decimal, hex, and binary with the numbers 0 to 15:

Decimal	Hex	Binary (Four Digit)
0	0	0000
1	1	0001
2	2	0010
3	3	0011
4	4	0100
5	5	0101
6	6	0110
7	7	0111
8	8	1000
9	9	1001
10	A	1010
11	B	1011
12	C	1100
13	D	1101
14	E	1110
15	F	1111

So, in hex, any **int** can be represented as a four-digit hex number. Thus, the binary number 10001100 would in hex be 8C. The C language has a special syntax for using hex numbers. You can assign a hex value to an **int** as follows:

```
int x = 0x8C;
```

As well as using hex notation for numbers, you can also use binary notation directly using the prefix "0b." For example, the binary used in the hex example of 0x8C could be written directly in binary as:

```
0b10001100
```

The Arduino standard library provides some functions that let you manipulate the 16 bits within an **int** individually. The function **bitRead** returns the value of a particular bit in an **int**; so, for the following example would assign the value 0 to the variable called **bit**:

```
int x = 0b10001100;
int bit = bitRead(x, 0);
```

In the second argument, the bit position starts at 0 and goes up to 15. It starts with the least significant bit. So the rightmost bit is bit 0, the next bit to the left is bit 1, and so on.

As you would expect, the counterpart to **bitRead** is **bitWrite**, which takes three arguments. The first is the number to manipulate, the second is the bit position, and the third is the bit value. The following example changes the value of the **int** from 2 to 3 (in decimal or hex):

```
int x = 0b10;
bitWrite(x, 0, 1);
```

Advanced I/O

There are some useful little functions that you can use to make your life easier when performing various input/output tasks.

Generating Tones

The **tone** function allows you to generate a square-wave signal (see Figure 7-3) on one of the digital output pins. The most common reason to do this is to generate an audible tone using a loudspeaker or buzzer.

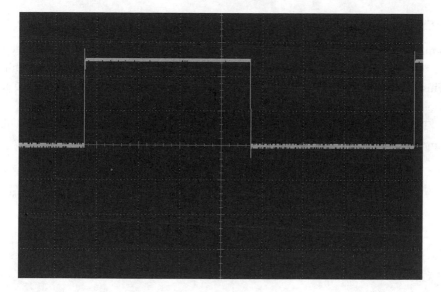

Figure 7-3   *A square-wave signal.*

The function takes either two or three arguments. The first argument is always the pin number on which the tone is to be generated, the second argument is the frequency of the tone in hertz (Hz), and the optional final argument is the duration of the tone. If no duration is specified, then the tone will continue playing indefinitely, as is the case in sketch 07_03_tone. This is why we have put the **tone** function call in **setup** rather than in the **loop** function.

```
//sketch 07_03_tone

void setup() {
   tone(4, 500);
}

void loop() {}
```

To stop a tone that is playing, you use the function **noTone**. This function has just one argument, which is the pin on which the tone is playing.

Feeding Shift Registers

Sometimes the Arduino Uno just doesn't have enough pins. When driving a large number of LEDs, for example, a common technique is to use a shift register chip. This chip reads data one bit at a time, and then when it has enough, it latches all those bits onto a set of outputs (one per bit).

To help you use this technique, there is a handy function called **shiftOut**. This function takes four arguments:

- The number of the pin on which the bit to be sent will appear.

- The number of the pin to be used as a clock pin. This toggles every time a bit is sent.

- A flag to determine whether the bits will be sent starting with the least significant bit or the most significant. This should be one of the constants **MSBFIRST** or **LSBFIRST**.

- The byte of data to be sent.

Interrupts

One of the things that tend to frustrate programmers used to "programming in the large" is that the Arduino can do only one thing at a time. If you like to have lots of threads of execution all running at the same time in your programs, then you are out of luck. Although a few people have developed projects that can execute multiple threads in this way, generally this capability is unnecessary for the type of uses that an Arduino is normally put to. The closest most Arduinos get to such execution is the use of interrupts.

Two of the pins on the Arduino Uno (D2 and D3) can have interrupts attached to them. That is, these pins act as inputs that, if the pins receive a signal in a specified way, the Arduino's processor will suspend whatever it was doing and run a function attached to that interrupt.

Sketch 07_04_interrupt blinks an LED, but then changes the blink period when an interrupt is received. You can simulate an interrupt by connecting your wire between pin D2 and GND and using the internal pull-up resistor to keep the interrupt high most of the time.

```
//sketch 07_04_interrupt
const int interruptPin = 2;
const int ledPin = 13;
int period = 500;
```

```
void setup() {
  pinMode(ledPin, OUTPUT);
  pinMode(interruptPin, INPUT_PULLUP);
  attachInterrupt(digitalPinToInterrupt(pin), goFast,
  FALLING);
}

void loop() {
  digitalWrite(ledPin, HIGH);
  delay(period);
  digitalWrite(ledPin, LOW);
  delay(period);
}

void goFast() {
  period = 100;
}
```

The following is the key line in the **setup** function of this sketch:

```
attachInterrupt(digitalPinToInterrupt(pin), goFast,
FALLING);
```

The first argument specifies which of the two interrupts you want to use. Rather confusingly, this is not just the pin name, to find the interrupt number, the function **digitalPinToInterrupt** is used.

Although the Arduino Uno only allows interrupts on two pins, some other boards allow interrupts on any of their pins.

The next argument is the name of the function that is to be called when there is an interrupt, and the final argument is a constant that will be one of **CHANGE, RISING,** or **FALLING.** Figure 7-4 summarizes these options.

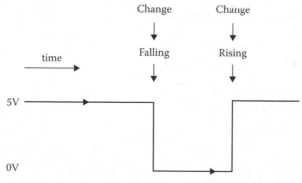

Figure 7-4 *Types of interrupt signals.*

If the interrupt mode is **CHANGE,** then either a **RISING** from 0 to 1 or a **FALLING** from 1 to 0 will both trigger an interrupt.

You can disable interrupts using the function **noInterrupts.** This stops all interrupts from both interrupt channels. You can resume using interrupts again by calling the function **interrupts.**

Different Arduino boards have different interrupt names for different pins so if you are not using an Arduino Uno, check the documentation for your board on http://www.arduino.cc.

Compile-Time Constants

Sometimes, when writing a sketch that might be used on different types of board, it can be useful for the sketch itself to be aware of the environment it will run on. For example, if the clock speed of the processor is below a certain threshold, you may want to disable some features of the sketch, because they would not work correctly. It may also be handy to know when (date and time) the sketch was flashed onto the board. This is possible using some special constants. These are summarized in the following table:

F_CPU	The CPU frequency of the board in MHz (16 for an Arduino Uno)
ARDUINO	The version of the Arduino IDE used to program the sketch onto the board
__DATE__	The date that the sketch was flashed onto the Arduino
__TIME__	The time that the sketch was flashed onto the Arduino

Sketch 07_05_compile_consts illustrates the use of these constants. If you run this sketch on an Arduino with the Serial Monitor open, you should see something like Figure 7-5.

```
//sketch 07_05_compile_consts
void setup() {
  Serial.begin(9600);
  Serial.println(__DATE__);
  Serial.println(__TIME__);
  Serial.println(F_CPU);
  Serial.println(ARDUINO);
}

void loop() {}
```

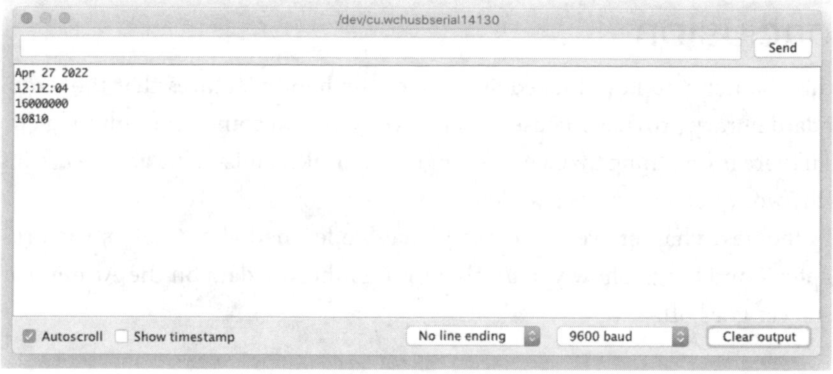

Figure 7-5 *Compile-time constants.*

The Arduino Web Editor

The Arduino WEB IDE is a browser-based version of the Arduino IDE (Figure 7-6). It has most of the same features as the regular IDE, and it allows you to keep your sketches safely stored in the cloud. It also integrates the Arduino documentation and provides the ability to deploy to some WiFi-enabled boards over the Internet, if you pay a small subscription. There is a free version to try, but you will have to sign up with an account at https://create.arduino.cc/editor and also install the Arduino Agent software onto your computer to allow access to the USB port to program your Arduino from the browser.

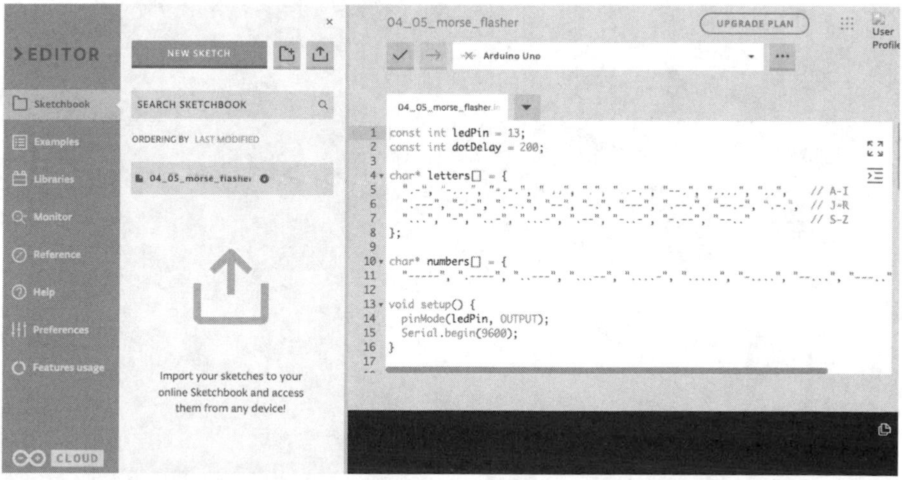

Figure 7-6 *The Arduino web IDE.*

Conclusion

In this chapter, you have looked at some of the handy features that the Arduino standard library provides. These features will save you some programming effort, and if there is one thing that a good programmer likes, it is being able to use high-quality work done by other people.

In the next chapter, we will extend what we learned about data structures in Chapter 4 and look at how you go about remembering data on the Arduino after the power goes off.

8

Data Storage

When you give values to variables, the Arduino board will remember those values only as long as the power is on. The moment that you turn the power off or reset the board, all that data is lost.

In this chapter, we look at some ways to hang on to that data either by storing variables in flash memory or on some Arduino boards electrically erasable programmable read-only memory (EEPROM).

Large Data Structures

Many of the latest Arduino-compatible boards such as the ESP32 and Pico boards have much more memory than you are ever likely to need. However, in boards like the Uno and Pro Mini, it is quite easy to run out of storage.

If the data that you want to store does not change, then you can just set the data up each time that the Arduino starts. An example of this approach is the case in the letters array in your Morse code translator of Chapter 4 (sketch 04_05_morse_flasher).

You used the following code to define a variable of the correct size and fill it with the data that you needed:

```
char* letters[] = {
  ".-", "-...", "-.-.", "-..", ".",
    "..-.", "--.", "....", "..",      // A-I
  ".---", "-.-", ".-..", "--", "-.",
    "---", ".--.", "--.-", ".-.",    // J-R
  "...", "-", "..-", "...-", ".--",
    "-..-", "-.--", "--.."           // S-Z
};
```

You may remember that you did the calculation and decided that you had plenty of your meager 2k (on an Arduino Uno) to spare. However, if you were using say an Uno and was a bit tight, you could store this data in the 32K of flash memory used to store programs, rather than the 2K of RAM. There is a means of doing this. It is a directive called **PROGMEM**; it lives in a library and is a bit awkward to use.

Storing Data in Flash Memory

This section only applies if you are using one of the "AVR" versions of Arduino such as the Uno, Nano, Mico, Pro Mini, or Leaonado. Arduinos and Arduino-compatibles based on newer processors have plenty of memory for storage, and in fact the method described here will not work on some of these boards.

To store your data in flash memory, you have to include the **PROGMEM** library as follows:

```
#include <avr/pgmspace.h>
```

The purpose of this command is to tell the compiler to use the **pgmspace** library for this sketch. In this case, a library is a set of functions that someone else has written and that you can use in your sketches without having to understand all the details of how those functions work.

Because you are using this library, the **PROGMEM** keyword and the **pgm_read_word** function are available. You will use both in the sketches that follow.

This library is included as part of the Arduino software and is an officially supported Arduino library. A good collection of such official libraries is available, and many unofficial libraries, developed by people like you and made for others to use, are also available on the Internet. Such unofficial libraries must be installed into your Arduino environment.

When using **PROGMEM**, you have to make sure that you use special **PROGMEM**-friendly data types. Unfortunately, that does *not* include an array of variable length **char** arrays. However, it does include access to an array of **char** arrays if those **char** arrays are of fixed size. The full program is very similar to that of sketch 04_05_morse_flasher in Chapter 4. You may like to open sketch 08_01_progmem in the IDE while I highlight the differences.

There is a new constant called maxLen that contains the maximum length of a single character's dots and dashes plus one for the null character on the end.

The structure to contain the letters now looks like this:

```
PROGMEM const char letters[26][maxLen] = {
    ".-", "-...", "-.-.", "-..", ".", "..-.", "--.", "....", "..",    // A-I
    ".---", "-.-", ".-..", "--", "-.", "---", ".--.", "--.-", ".-.",  // J-R
    "...", "-", "..-", "...-", ".--", "-..-", "-.--", "--.."          // S-Z
};
```

The **PROGMEM** keyword indicates that the data structure is to be stored in flash. You can only store constants like this; once in the flash, the data structure cannot be changed, hence the use of **const**. The size of the array also has to be fully specified as 26 letters by **maxLen** (minus 1) dots and dashes.

The loop function is also slightly different from the original sketch.

```
void loop() {
  char ch;
  char sequence[maxLen];
  if (Serial.available() > 0) {
    ch = Serial.read();
    if (ch >= 'a' && ch <= 'z') {
      memcpy_P(&sequence, letters[ch - 'a'], maxLen);
      flashSequence(sequence);
    }
    else if (ch >= 'A' && ch <= 'Z') {
      memcpy_P(&sequence, letters[ch - 'A'], maxLen);
      flashSequence(sequence);
    }
    else if (ch >= '0' && ch <= '9') {
      memcpy_P(&sequence, numbers[ch - '0'], maxLen);
      flashSequence(sequence);
    }
    else if (ch == ' ') {
      delay(dotDelay * 4);   // gap between words
    }
  }
}
```

The data may look like an array of strings, but actually internally it is stored in flash in a way that can only be accessed by the special function **memcp_P**, which copies the flash data into a **char** array called **sequence** that is initialized to **maxSize** characters in length.

The & character before **sequence** allows **memcpy_P** to modify the data inside the sequence character array.

I have not listed sketch 08_01_progmem here, as it is a little lengthy, but you may wish to load it and verify that it works the same way as the RAM-based version.

In addition to creating the data in a special way, you also have to read the data back a special way. Your code to get the code string for a Morse letter from the array has to be modified to look like this:

```
strcpy_P(buffer, (char*)pgm_read_word(&(letters[ch - 'a'])));
```

This uses a **buffer** variable into which the **PROGMEM** string is copied, so that it can be used as a regular **char** array. This needs to be defined as a global variable as follows:

```
char buffer[6];
```

This approach works only if the data is constant—that is, you are not going to change it while the sketch is running. In the next section, you will learn about using the EEPROM that is intended for storing persistent data that can be changed.

If you have individual strings that are perhaps formatted for messages to be displayed on the serial monitor, then Arduino C provides a handy shortcut. You can just enclose the string in F() as shown in this example:

```
Serial.println(F("Hello World"));
```

The string will then be stored in flash memory, rather than use up RAM.

EEPROM

EEPROM is a type of memory whose contents are retrained even when the power is removed.

The ATMega328 at the heart of an Arduino Uno has a kilobyte of EEPROM. EEPROM is designed to remember its contents for many years. Despite its name, it is not really read-only. You can write to it.

The official Arduino commands for reading and writing to EEPROM are just as awkward to use as the ones for using **PROGMEM**. You have to read and write to and from EEPROM one byte at a time.

The example of sketch 08_02_eeprom_byte allows you to enter a number between 0 and 255 from the Serial Monitor. The sketch then remembers the number and repeatedly writes it out on the Serial Monitor.

```
//sketch 08_02_eeprom_byte
#include <EEPROM.h>

int addr = 0;
byte b;

void setup() {
  Serial.begin(9600);
  b = EEPROM.read(addr);
}

void loop() {
  if (Serial.available() > 0) {
    int i = Serial.parseInt();
    if (i <= 255) {
      b = lowByte(i);
      EEPROM.write(addr, b);
      Serial.println("saved byte");
    }
    else {
      Serial.println("number too big");
    }
  }
  Serial.println(b);
  delay(1000);
}
```

To try this sketch, open the Serial Monitor and enter a number between 0 and 255. Then unplug the Arduino and plug it back in. When you reopen the Serial Monitor, you will see that the number has been remembered.

The function **EEPROM.write** takes two arguments. The first is the address, which is the memory location in EEPROM and should be between 0 and 1023. The second argument is the data to write at that location. This must be a single byte.

Storing an int in EEPROM

An **int** requires 2 bytes of storage. To store a two-byte **int** in locations 0 and 1 of the EEPROM, you could do this:

```
int x = 1234;
EEPROM.write(0, highByte(x));
EEPROM.write(1, lowByte(x));
```

The functions **highByte** and **lowByte** are useful for separating an **int** into two bytes. Figure 8-1 shows how this **int** is actually stored in the EEPROM.

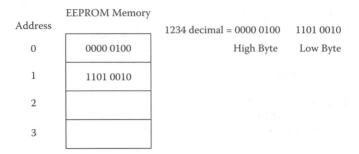

Figure 8-1 *Storing a 16-bit integer in EEPROM.*

To read the **int** back out of EEPROM, you need to read the two bytes from the EEPROM and reconstruct the **int**, as follows:

```
byte high = EEPROM.read(0);
byte low = EEPROM.read(1);
int x = (high << 8) + low;
```

The << operator is a bit shift operator that moves the eight high bytes to the top of the **int** and then adds in the low byte.

You can find an example sketch for this in sketch 08_03_eeprom_int. This works rather like its counterpart for bytes, but allows you to enter the whole range of **int** numbers from −32768 to 32767.

```
//sketch 08_03_eeprom_int
#include <EEPROM.h>

int i;

void setup() {
  Serial.begin(9600);
  i = readEEPROMint(0);
}

void writeEEPROMint(int addr, int x) {
  EEPROM.write(addr, highByte(x));
  EEPROM.write(addr + 1, lowByte(x));
}

int readEEPROMint(int addr) {
  int high = EEPROM.read(addr);
  int low = EEPROM.read(addr + 1);
  return (high << 8) + low;
}
```

```
void loop() {
  if (Serial.available() > 0) {
    i = Serial.parseInt();
    writeEEPROMint(0, i);
    Serial.println("saved int");
  }
  Serial.println(i);
  delay(1000);
}
```

Writing Anything to EEPROM

There is a neat way of saving and reading any variable to and from EEPROM. This uses a C++ technique called generics to make a pair of general purpose functions that can save and read different types of data to EEPROM.

To use this, just include the two functions in your code. As an example sketch 08_04_eeprom_long saves a long value (4 bytes).

```
//sketch 08_04_eeprom_long
#include <EEPROM.h>

long x = 12345678;
long y = 0;

void setup() {
  Serial.begin(9600);
  EEPROM_writeAnything(0, x);
  Serial.print("wrote x: ");
  Serial.println(x);
  int n = EEPROM_readAnything(0, y);
  Serial.print("read y: ");
  Serial.println(y);
  Serial.println(n);
}

void loop() {}

template <class T> int EEPROM_writeAnything(int ee,
  const T& value) {
  const byte* p = (const byte*)(const void*)&value;
  int i;
  for (i = 0; i < sizeof(value); i++) {
    EEPROM.write(ee++, *p++);
  }
  return i;
}
```

```
template <class T> int EEPROM_readAnything(int ee,
  T& value) {
    byte* p = (byte*)(void*)&value;
    int i;
    for (i = 0; i < sizeof(value); i++) {
      *p++ = EEPROM.read(ee++);
    }
    return i;
}
```

The functions EEPROM_writeAnything and EEPROM_readAnything both return the size of the data that was saved in bytes. When you open the serial console you will see output something like this:

wrote x: 12345678

read y: 12345678

4

The long variable x is saved and its result read back into the variable y. The 4 in the output indicates that 4 bytes were used.

Storing a float in EEPROM

Storing a float in EEPROM using the EEPROM_writeAnything function is very similar to storing an **int** as sketch 08_05_eeprom_float illustrates.

```
//sketch 08_05_eeprom_float
#include <EEPROM.h>

float x = 12.34;
float y = 0;

void setup() {
  Serial.begin(9600);
  EEPROM_writeAnything(0, x);
  Serial.print("wrote x: ");
  Serial.println(x);
  int n = EEPROM_readAnything(0, y);
  Serial.print("read y: ");
  Serial.println(y);
  Serial.println(n);
}
```

Note that the float data type is also 4 bytes long.

Storing a String in EEPROM

Writing and reading character strings into the EEPROM is also best accomplished using EEPROM_writeAnything. Sketch 08_06_eeprom_string illustrates this with an example that reads and writes passwords from EEPROM. The sketch first displays this password read from EEPROM and then prompts you to enter a new password (Figure 8-2). Having set the password, you can unplug the Arduino to power it down and when you plug it back in again and open the Serial Monitor, the old password will still be there.

```
//sketch 08_06_eeprom_string
#include <EEPROM.h>
const int maxPasswordSize = 20;
char password[maxPasswordSize];

void setup() {
  EEPROM_readAnything(0, password);
  Serial.begin(9600);
}

void loop() {
  Serial.print("Your password is:");
  Serial.println(password);
  Serial.println("Enter a NEW password");
  while (!Serial.available()) {};
  int n = Serial.readBytesUntil('\n', password,
          maxPasswordSize);
  password[n] = '\0';
  EEPROM_writeAnything(0, password);
  Serial.print("Saved Password: ");
  Serial.println(password);
}
```

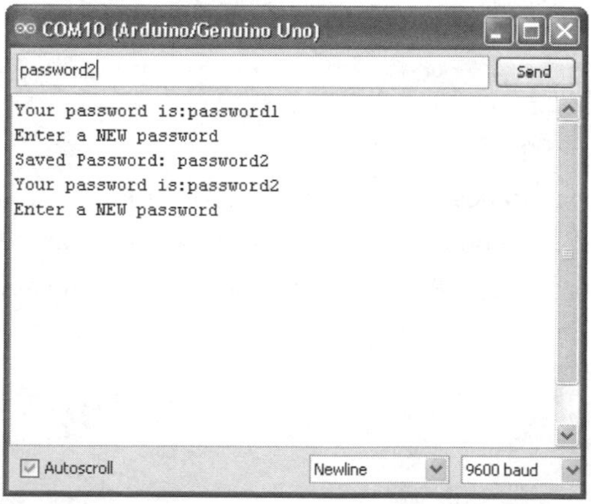

Figure 8-2 *Missing figure caption.*

The character array **password** has a fixed size of 20 characters that must also include the "\0" end marker. In the **startup** function the contents of EEPROM starting at location 0 are read into **password**.

The **loop** function displays the necessary messages and then the **while** loop does nothing until serial communication arrives, indicated by **Serial.available** returning more than 0. The **readBytesUntil** function will then keep reading characters until the end of line character "\n" is encountered. The bytes being read will be put straight into the **password char** array.

Because you don't know how long a password will be entered, the result of reading the bytes is stored in **n** and then element **n** of the password is set to '\0' to mark the end of the string. Finally, the new password is printed to the Serial Monitor to confirm the change in password.

Clearing the Contents of EEPROM

When writing to EEPROM, remember that even uploading a new sketch will not clear the EEPROM, so you may have leftover values in there from a previous project. Sketch 08_07_eeprom_clear resets all the contents of EEPROM to zeros:

```
//sketch 08_07_eeprom_clear
#include <EEPROM.h>

void setup() {
  Serial.begin(9600);
  Serial.println("Clearing EEPROM");
  for (int i = 0; i < 1024; i++) {
    EEPROM.write(i, 0);
  }
  Serial.println("EEPROM Cleared");
}

void loop() {}
```

Also be aware that you can write to an EEPROM location only about 100,000 times before it will become unreliable. So only write a value back to EEPROM when you really need to. EEPROM is also quite slow, taking about 3 milliseconds to write a byte.

Compression

When saving data to EEPROM or when using **PROGMEM**, you will some-times find that you have more to save than you have room to save it. When this happens, it is worth finding the most efficient way of representing the data.

Range Compression

You may have a value for which on the face of it you need an **int** or a **float** that is 16-bit. For example, to represent a temperature in degrees Celsius, you might use a **float** value such as 20.25. When you are storing that into EEPROM, life would be so much easier if you could fit it into a single byte, and you could store twice as much as if you used a **float**.

One way that you can do this is to change the data before you store it. Remember that a byte will allow you to store a positive number between 0 and 255. So if you only cared about the temperature to the nearest degree Celsius, then you could simply convert the **float** to an **int** and discard the part after the decimal point. The following example shows how to do this:

```
int tempInt = (int)tempFloat;
```

The variable **tempFloat** contains the floating point value. The **(int)** command is called a *type cast* and is used to convert a variable from one type to another compatible type. In this case, the type cast converts the **float** of (for example) 20.25 to an **int** that will simply truncate the number to 20.

If you know that the highest temperature that you care about is 60 degrees Celsius and that the lowest is 0 degrees Celsius, then you could multiply every temperature by 4 before converting it to a byte and saving it. Then when you read the data back from EEPROM, you can divide by 4 to get a value that has a precision of 0.25 of a degree.

The following code example (sketch 08_08_eeprom_compress) saves such a temperature into EEPROM, then reads it back and displays it in the Serial Monitor as proof:

```
//sketch 08_08_eeprom_compress
#include <EEPROM.h>
```

```
void setup() {
  float tempFloat = 20.75;
  byte tempByte = (int)(tempFloat * 4);
  EEPROM.write(0, tempByte);

  byte tempByte2 = EEPROM.read(0);
  float temp2 = (float)(tempByte2) / 4;
  Serial.begin(9600);
  Serial.println("\n\n\n");
  Serial.println(temp2);
}

void loop(){}
```

There are other means of compressing data. For instance, if you are taking readings that change slowly—again, changes in temperature are a good example of this—then you can record the first temperature at full resolution and then just record the changes in temperature from the previous reading. This change will generally be small and occupy fewer bytes.

Conclusion

You now know a little about how to make your data hang around after the power has gone off. In the next chapter, you will look at displays.

9

Displays

In this chapter, you look at how to write software to control displays. Figure 9-1 shows the two types of display that you will use. The first is an alphanumeric liquid crystal display (LCD) display shield. The second is a 128 × 64-pixel OLED (organic light-emitting diode) graphical display. These two types of display are very popular for the Arduino.

This is a book about software, not hardware; but in this chapter, we will have to explain a little about how the electronics of these displays work so that you understand how to drive them.

Figure 9-1 *An alphanumeric LCD shield (left) and OLED display (right).*

Alphanumeric LCD Displays

The LCD module that we use is an Arduino shield that can just be plugged on top of an Arduino Uno shaped board. In addition to its display, it also has some buttons. There are a number of different shields, but nearly all of them use the same LCD controller chip (the HD44780), so look for a shield that uses this controller chip.

I used the DFRobot LCD Keypad Shield for Arduino. This module supplied by DFRobot (www.robotshop.com) is inexpensive and provides an LCD display that is 16 characters by 2 rows and also has 6 pushbuttons.

The shield comes assembled, so no soldering is required; you just plug it on top of your Arduino Uno board (see Figure 9-2).

The LCD shield uses seven of the Arduino pins to control the LCD display and one analog pin for the buttons. So we cannot use these Arduino pins for any other purpose.

Figure 9-2 *LCD shield attached to an Arduino board.*

A USB Message Board

For a simple example of a simple use of the display, we are going to make a USB message board. This will display messages sent from the Serial Monitor.

The Arduino IDE comes with an LCD library. This greatly simplifies the process of using an LCD display. The library gives you useful functions that you can call:

- **clear** clears the display of any text.

- **setCursor** sets the position in row and column where the next thing that you print will appear.

- **print** writes a string at that position.

This example is listed in sketch 09_01_message_board:

```
//sketch 09_01_message_board
#include <LiquidCrystal.h>

// lcd(RS E D4 D5 D6 D7)
LiquidCrystal lcd(8, 9, 4, 5, 6, 7);
int numRows = 2;
int numCols = 16;

void setup() {
  Serial.begin(9600);
  lcd.begin(numRows, numCols);
  lcd.clear();
  lcd.setCursor(0,0);
  lcd.print("Arduino");
  lcd.setCursor(0,1);
  lcd.print("Rules");
}

void loop()
{
  if (Serial.available() > 0) {
    char ch = Serial.read();
    if (ch == '#') {
      lcd.clear();
    }
```

```
    else if (ch == '/') {
       // new line
       lcd.setCursor(0, 1);
    }
    else {
       lcd.write(ch);
    }
  }
}
```

As with all Arduino libraries, you have to start by including the library to make the compiler aware of it.

The next line defines which Arduino pins are used by the shield and for what purpose. If you are using a different shield, then you may well find that the pin allocations are different, so check in the documentation for the shield.

In this case, the six pins used to control the display are D4, D5, D6, D7, D8, and D9. The purpose of each of these pins is described in Table 9-1.

The **setup** function is straightforward. You start serial communications so that the Serial Monitor can send commands and initialize the LCD library with the dimensions of the display being used. You also display the message

Parameter to LCD()	Arduino Pin	Purpose
RS	8	Register Select; this is set to 1 or 0 depending on whether the Arduino is sending data for characters or an instruction. An instruction might make the cursor flash, for example.
E	9	Enable; this gets toggled to tell the LCD controller chip that the data on the following four pins is ready to be read.
Data 4	4	These four pins are used to transfer data. The LCD controller chip used by the shield can use eight-bit or four-bit data. This shield uses four bits, in which case the bits 4–7 are used rather than 0–7.
Data 5	5	
Data 6	6	
Data 7	7	

Table 9-1 *LCD Shield Pin Assignments*

"Arduino Rules" on two lines by setting the cursor to top-left, printing "Arduino," then moving the cursor to the start of the second row and printing "Rules."

Most of the action takes place in the **loop** function, which checks for any incoming characters from the Serial Monitor. The sketch deals with characters one at a time.

Apart from ordinary characters that the sketch will simply display, there are also a couple of special characters. If the character is a #, then the sketch clears the whole display, and if the character is a /, the sketch moves to the second line. Otherwise, the sketch simply displays the character at the current cursor position using **write**. The function **write** is like **print**, but it prints only a single character rather than a string of characters.

Using the Display

Try out sketch 09_01_message_board by uploading it to the board and then attaching the shield. Note that you should always unplug the Arduino board so that it is off before you plug in a shield.

Open up the Serial Monitor and try typing in the text shown in Figure 9-3.

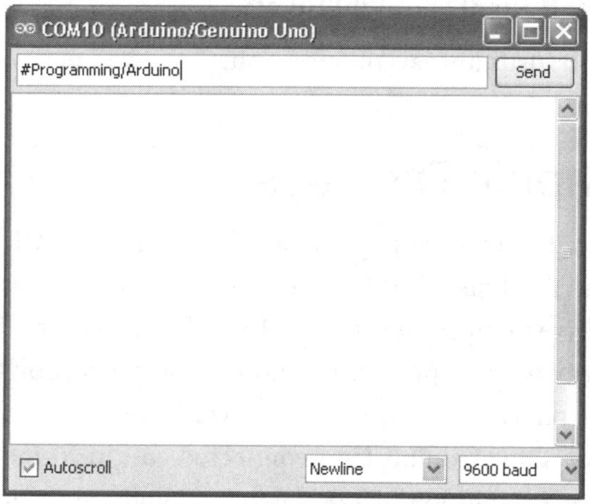

Figure 9-3 *Sending commands to the display.*

Other LCD Library Functions

In addition to the functions that you have used in this example, there are a number of other functions that you can use:

- **home** is the same as **setCursor(0,0)**: it moves the cursor to top-left.

- **cursor** displays a cursor.

- **noCursor** specifies not to display a cursor.

- **blink** makes the cursor blink.

- **noBlink** stops the cursor from blinking.

- **noDisplay** turns off the display without removing the content.

- **display** turns the display back on after **noDisplay**.

- **scrollDisplayLeft** moves all the text on the display one character position to the left.

- **scrollDisplayRight** moves all the text on the display one character position to the right.

- **autoscroll** activates a mode in which, as new characters are added at the cursor, the existing text is pushed in the direction determined by the functions **leftToRight** and **rightToLeft.**

- **noAutoscroll** turns **autoscroll** mode off.

OLED Graphic Displays

OLED displays are bright and crisp and are fast replacing LCD displays in consumer appliances. The type of OLED described here uses an interface bus called I2C and has a driver chip called the SD1306. These can be bought on eBay, Adafruit, and many other suppliers around the Internet. Look for a device with just four interface pins as these are easiest to work with.

Figure 9-4 shows an Arduino Uno connected to a 0.96-inch OLED display. These boards have a resolution of 128 × 64 pixels and are monochrome—in this case, blue. The popularity of these boards mean that the Arduino community has ported the code to drive these displays to most Arduino-compatible boards.

Figure 9-4   *An Arduino Uno and OLED display.*

Connecting an OLED Display

You can connect your OLED display to your Arduino using jumper wires. You can buy these from many sources including Adafruit and depending on the type of Arduino, or Arduino-compatible board, you will either need female-to-male or female-to-female jumper wires. You will need to make the following connections:

- GND on the display to GND on the Arduino

- VCC on the display to 5V on the Arduino

- SCL on the display to the SCL pin of the Arduino. These are labelled on the underside of an Arduino Uno and are also indicated in Figure 9-5

- SDA on the display to the SDA pin of the Arduino (also see Figure 9-5)

I2C (pronounced *I squared C*) is a serial bus standard commonly used to connect sensors and displays to microcontrollers like the Arduino. As well as the ground (GND) and positive power pins, it uses a data pin (SDA) and clock pin (SCK) to communicate with the microcontroller by sending serial data 1 bit at a time.

SCL SDA

Figure 9-5 *Identifying the SCL and SDA pins of an Arduino Uno.*

Software

Sketch 09_02_oled will count in seconds up to 9999 and then reset to 0.

Before uploading it to your Arduino, you need to find out the I2C address of the display. This will be a hexadecimal number and may be written on the back of the OLED display. Many low-cost eBay OLED displays use 0x3c.

You will also need to install some libraries before the sketch will compile. These can both be imported directly from the Arduino IDE's Library Manager. Open the Library Manager by selecting the menu option Sketch | Include Library | Manage Libraries…. Then search for "Adafruit GFX Library" and click Install (Figure 9-6). Then do the same for the "Adafruit SSD1306" library. The SPI and Wire libraries that the sketch needs are both installed by default in the Arduino IDE.

```
//sketch 09_02_oled
#include <SPI.h>
#include <Wire.h>
#include <Adafruit_GFX.h>
#include <Adafruit_SSD1306.h>

Adafruit_SSD1306 display(128, 64, &Wire, -1);
```

```
void setup() {
  display.begin(SSD1306_SWITCHCAPVCC, 0x3c);  // may need to change this
  display.setTextSize(4);
  display.setTextColor(WHITE);
}

void loop() {
  static int count = 0;
  display.clearDisplay();
  display.drawRoundRect(0, 0, 127, 63, 8, WHITE);
  display.setCursor(20,20);
  display.print(count);
  display.display();
  count ++;
  if (count > 9999) {
    count = 0;
  }
  delay(1000);
}
```

The sketch starts by importing the libraries that it needs and then a **display** variable initialized. The parameter supplied is that of the optional "reset" pin that some OLED displays (including those supplied by Adafruit) have. If your display does not have a reset pin, then set this value to -1.

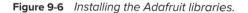

Figure 9-6   *Installing the Adafruit libraries.*

The **setup** function initializes the display and you may need to change the I2C address supplied as the second parameter from 0x3c to a different value. **setup** then sets the font size to 4 (large) and the text color to white (anything but black will display in the LED color).

The **loop** function clears the display, draws a round-cornered rectangle, sets the cursor position, and then prints the value of count. The display will not actually be updated until the command **display.display()** is run. The variable **count** is then incremented and there is a delay of one second.

The Adafruit GFX library provides all sorts of fancy drawing routines that you can use with the graphical display. For documentation on this library see https://learn.adafruit.com/adafruit-gfx-graphics-library.

Some boards such as the ESP32 boards are capable of using I2C on any of their GPIO pins. For this, an extra line of code is needed in setup to specify which pins are to be used. You can find an example of this in the sketch 09_03_oled_esp32. The new line of code looks like this:

```
Wire.begin(17, 16); // SDA, SCL
```

Conclusion

You can see that programming displays is not hard, particularly when there is a library that can do a lot of the work for you.

In the next chapter, you will use an Arduino to connect to your network and the Internet.

10

Arduino Internet of Things Programming

The Internet of Things (IoT) is the concept that more and more devices will become connected to the Internet. That doesn't just mean more and more computers using browsers, but actual appliances and wearable and portable technology. This includes all sorts of home automation from smart appliances and lighting, to security systems and even Internet-operated pet feeders as well as lots of less practical but fun projects.

In this chapter, you will learn how to program a WiFi-capable board to both send web requests to services on the Internet and program the device to act as a web server on your local network.

For this chapter, you will need a WiFi-capable board such as the widely available Lolin32 Lite.

Boards for IoT

The Arduino Uno is a great board for getting started with Arduino; however, it lacks WiFi hardware and although it can be kitted out with a WiFi shield, this makes an expensive and bulky option for an IoT project. It is far more practical to use an ESP32-based board like the Lolin32 shown in Figure 10-1.

ESP32 boards come in many shapes and sizes and some have extra features such as a battery connector, to allow the board to charge a LiPo battery from universal serial bus (USB) and then use it to power the board when it is deployed into a project. You can even find ESP32 boards with a tiny organic light emitting diode (OLED) display like the display described in Chapter 9 or even a tiny camera.

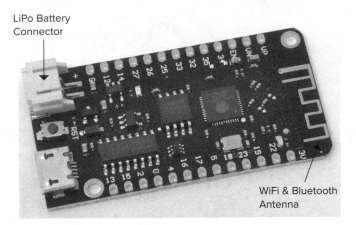

LiPo Battery
Connector

WiFi & Bluetooth
Antenna

Figure 10-1 *A Lolin32 Lite ESP32 board.*

Installing ESP32 into the Arduino IDE

We first met the ESP32-type board in Chapter 6, where you will find instructions for adding support to the ESP32-type boards to the Arduino IDE. Having installed support, select the board type to match your board (Figure 10-2).

You will also need to set the port, and if you have trouble uploading, try decreasing the upload speed in the Tools menu. Before launching into IoT code, try uploading a simple blink sketch such as sketch 02_01_blink or the sketch below. Remember to change the pin to blink to be the pin that the board's built-in LED is connected to. You can use the constant LED_BUILTIN to do this.

```
void setup() {
  pinMode(LED_BUILTIN, OUTPUT);
}

void loop() {
 digitalWrite(LED_BUILTIN, HIGH);
 delay(500);
 digitalWrite(LED_BUILTIN, LOW);
 delay(500);
}
```

If you have been using an Arduino Uno, then the first thing you will notice about the ESP32 boards is that compiling and uploading takes a lot longer.

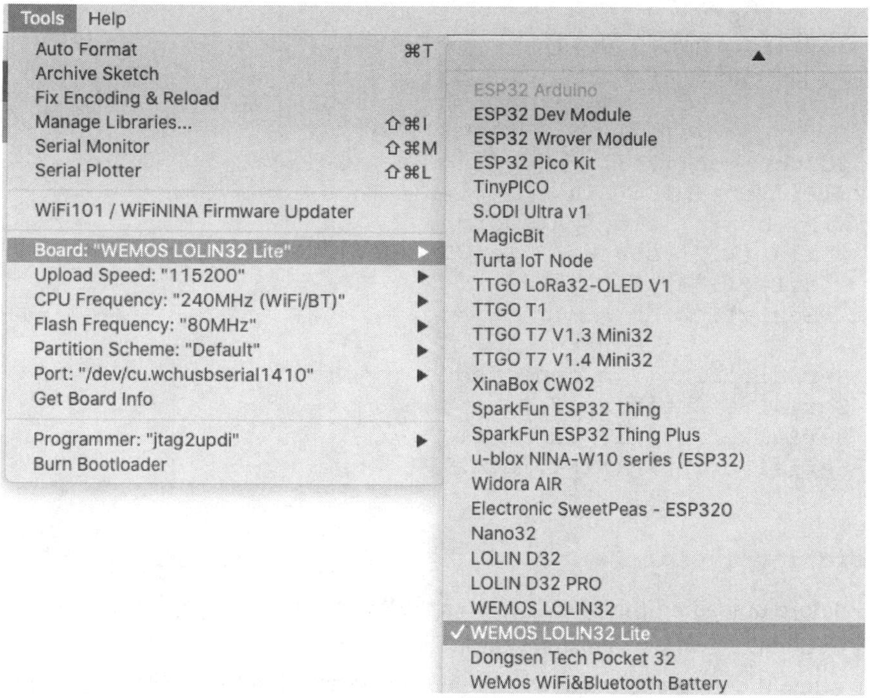

Figure 10-2 *Setting the board type.*

Connecting to WiFi

Although Ethernet shields are available for Arduinos, it is usually more convenient to connect wirelessly to your network using WiFi. This is the same process that you go through when you connect a smartphone or computer to your home WiFi for the first time. You will supply the board with the name of your network (called the SSID) and the password. On a phone or computer, you would choose the SSID from a list and then enter the password when prompted. When connecting a board programmed using the Arduino IDE, you put this information in the sketch. In sketch 10_01_wifi_connect, we just go as far as connecting to the network, but this will form the bases of all the other network examples in this chapter.

```
//sketch 10_01_wifi_connect
#include <WiFi.h>

// Change these 2 for your network!
const char* ssid = "my network name";
const char* password = "my password";
```

```
void setup(void) {
  Serial.begin(9600);
  connectWiFi();
}

void connectWiFi() {
  WiFi.mode(WIFI_STA);
  WiFi.begin(ssid, password);
  while (WiFi.status() != WL_CONNECTED) {
    delay(500);
    Serial.print(".");
  }
  Serial.print("\nConnected to: ");
  Serial.println(ssid);
  Serial.print("IP address: ");
  Serial.println(WiFi.localIP());
}

void loop(void) {}
```

Before uploading this sketch to your ESP32 board, change the two lines for ssid and password to match your WiFi credentials.

When the program is uploaded, open the serial monitor so that you can see what's happening. You should see something like this:

```
. . . . .
Connected to: MY_NETWORK
IP address: 192.168.1.229
```

While the board is connecting to your WiFi network, you will see a series of dots appear. After a few seconds, the board will confirm that it is connected and report the IP address of your board on your local network. Every device that connects to your network is allocated an IP address by your network. You can think of this as your devices name on the network, and it's that that is used whenever you need to communicate with the device.

Starting at the top of the code, we not unsurprisingly, start by importing the WiFi library. All the code for actually connecting to the network is contained in the function connectWiFi. This starts by setting the WiFi mode to STA (for station). This means that the board is going to connect to an existing network. The WiFi.begin call starts of the process of connecting to the network. As this takes a few seconds, the next few lines monitor the status of WiFi and print out dots until connection has taken place. After that the code just sends messages to the serial monitor confirming successful connection and the IP address allocated to the board by your router.

Running a Web Server

Let's now take the connection example a stage further and have our ESP32 board act as a web server. Yes, I did just say that. This little board is going to do in miniature the same thing as a big server on the Internet. However, as you might expect, this server is not going to be able to handle thousands of connections at a time (actually, it just handles 1). The other difference is that this server is only going to be available inside our local network. It is just for us, the world is not going to have access to it.

When we connect to this server using a browser on our computer or phone, all it's going to do is display a message to us in the browser like the one in Figure 10-3.

If we really wanted a proper web server, then there are much better choices of hardware available. In this case, we are just displaying a message, but you could imagine a situation where the ESP32 board was connected to some sensors (perhaps a weather station) and then it could serve up its readings, so that you could view them from your phone or computer.

Here is the code for sketch 10_02_webserver_hello. This uses the same function `connectWiFi` as the previous sketch; so, for brevity, that function has been omitted from the following listing as has the `loop` function which hasn't changed either.

```
//sketch 10_02_webserver_hello
#include <WiFi.h>
#include <WebServer.h>
#include <ESPmDNS.h>

// Change these 2 for your network!
const char* ssid = "my network name";
const char* password = "my password";
const char* hostname = "esp32";

WebServer server(80);

void handleRoot() {
  server.send(200, "text/html", "<h1>Hello World!</h1>");
}

void setup(void) {
  Serial.begin(9600);
  connectWiFi();
  if (MDNS.begin(hostname)) {
    Serial.println("Webserver started");
  }
```

```
server.on("/", handleRoot);
server.begin();
Serial.print("Open your browser on http://");
Serial.println(WiFi.localIP());
Serial.print("this may also work: http://");
Serial.print(hostname); Serial.println(".local");
}
```

There are a couple of new imports. The purpose of WebServer.h should be fairly obvious. The other import (ESPmDNS.h) provides a feature called mDNS (multicast Domain Name Service). This allows devices on the network to identify themselves by name in addition to their IP address. The new constant hostname is used by mDNS.

The line WebServer server(80); sets up a web server running on port 80 (the default for all web servers). The WebServer library is designed to work like most web servers by being able to serve up pages when a browser demands them. The default page is the root or index page and is the page that you see when the URL is just the host, without specifying a page—for example http://192.168.1.229 shown in Figure 10-3.

The function handleRoot will be called whenever a browser requests the root page. This function responds to the browser by sending the code 200 (no errors or redirects), content type of text/html and the HTML tag <h1>Hello World!</h1>. The h1 tag means heading level 1, which is why the writing is big when you see it in the browser.

In the setup function MDNS.begin registers the hostname with mDNS. The mDNS may or may not work, depending on your network and the computer you are using to try and connect to the ESP32 web server, but the IP address should always work. If you are happy just using the IP address, then you can omit the whole if statement.

The code server.on("/", handleRoot); is what links requests to the server for the root page to the handleRoot function.

Figure 10-3 *Hello Web Server!*

To try the sketch out, don't forget to change the `ssid` and `password` to match your network and then copy and paste the URL from the serial monitor into the address bar of your browser.

Serving Sensor Readings

A more realistic thing for our web server to do would be to report sensor readings from the ESP32 board. The ESP32 has certain pins that are touch sensitive. When you touch them a reading decreases. The lower the number, the better the touch. We can use this facility to provide a value to be displayed on our web server. We will start with a really simple version of this example, where the readings only update when you refresh the page and then go on to improve the example so that the readings update on the webpage automatically.

The code for this sketch is in 10_03_webserver_touch. The only part of the sketch that has changed is the function `handleRoot` and a new constant `touchPin`.

```
//sketch 10_03_webserver_touch
void handleRoot() {
  String message = "<h1>Touch value: ";
  message += touchRead(touchPin);
  message += "</h1>";
  server.send(200, "text/html", message);
}
```

The `handleRoot` function now builds up a message to be sent to the browser that is still a level 1 heading (h1) but now the text will display the value of the pin being used to sense touch, defined in the constant `touchPin`.

Try uploading the sketch (don't forget to change the SSID and password) and you should see something like Figure 10-4.

Put your finger on pin 13 of your ESP board and at the same time click on the reload button on your browser. The reading should change to a much lower value.

Touch value: 58

Figure 10-4   *Serving touch readings from an ESP32 web server.*

Serving Sensor Readings—Improved

Having to click on the reload button in your browser to get updated readings is a bit crude. The way to fix this, so that the readings update automatically, is to add some Javascript to the root page that tells the browser how to fetch values to be displayed periodically and then for the browser to update the page it's displaying its self.

This is where it can get very confusing as to where the code we see in the Arduino sketch is actually running. This is further complicated by having some of the code being use in a different programming language (JavaScript) because that's the programming language that browsers use).

We are going to change our web server code as follows:

- Put the contents of the root page into a separate file. This root page will contain not only the HTML tags to display the reading in, but also, crucially, this page will also provide the JavaScript code to the browser that it will need to run to update the page.

- Add a second page (called *touch*) to the web server that responds with a simple text value that is the touch reading.

A web browser pointed to the root page will only need to load the page contents once and the browser displaying the page will repeatedly request reading every half a second from the touch page.

When you upload this sketch, it will look the same as Figure 10-4 but importantly, you should see the touch reading automatically update as you touch or release the pin, without having to reload the page.

Most of this code has already been explained. Let's look at the new bits.

First of all, when you open this sketch, you will see that there are now two tabs in the Arduino IDE's editor window. The usual one being labeled 10_04_web-server_touch_auto and the new one being labeled index.h. Arduino sketches are usually small enough that there is no need to split things out into separate files. However, this is something that is useful to do in situations like this. To add a new file to your sketch, click on the drop-down icon at the far right of the tab bar (Figure 10-5) and select New Tab. You will then be prompted to enter a file name.

Here's the contents of index.h.

```
char *index_template = "                                    \
<script>                                                     \
```

```
function get_reading() {                                         \
  const request = new XMLHttpRequest();                          \
  request.open('GET', '/touch');                                 \
  request.send();                                                \
  request.onload = function() {                                  \
    if (request.status === 200) {                                \
      value_got = request.responseText;                          \
      field = document.getElementById('value_field');  \
      field.textContent = value_got;                             \
      window.setTimeout(get_reading, 500);                       \
    }                                                            \
  }                                                              \
}                                                                \
get_reading();                                                   \
                                                                 \
                                                                 \
</script>                                                        \
<h1>Touch Value: <span id='value_field'/></h1>                   \
";
```

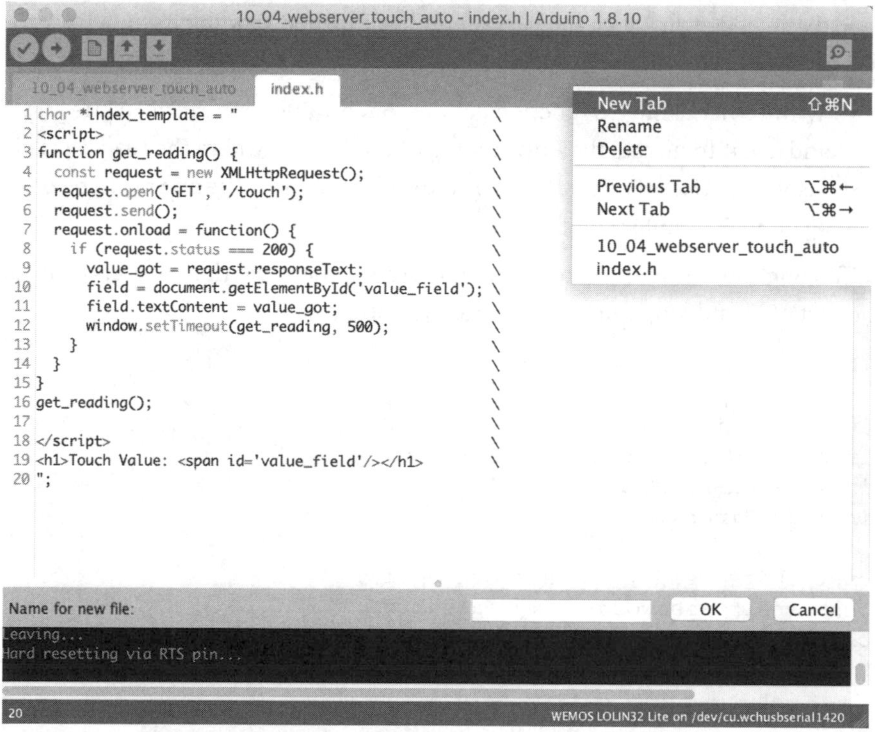

Figure 10-5   *Adding a file to a sketch.*

The file is actually legal Arduino C code defining a character string and assigning it to a constant called `index_template`. The \ characters on the right of each line are string continuation characters that allow the string to be spread over multiple lines to make it easier to read.

It is beyond the scope of this book to teach JavaScript in any detail, but in explaining how this example works you shouldn't find it too hard to adapt it to other uses. The contents of index.h are really in two parts: Javascript code contained inside the `<script>` tag and then some regular HTML inside an `<h1>` tag. Remember this JavaScript code is not going to run on the ESP32 board. The ESP32 board is just sending it to the browser so that the browser can then run it. The JavaScript code first defines a function (just like an Arduino C function) called `get_reading` and then calls `get_reading` just once. Here's what `get_reading` does:

- Define a web request of type 'get' (get a value) to access the page 'touch' and then send it to the web server.

- Define a new nameless function and associate it with `request.onload` that will be run, when the web server has finished sending its data back to the browser.

- Within that nameless function get the touch value sent back (`value_got`) and use it to update the `<span>` tag of the HTML where the reading is displayed. Use `window.timeout` to schedule `get_reading` to be called again in 500 milliseconds.

Turning our attention to the main sketch, here is the listing, with the connectWiFi and loop functions omitted for brevity.

```
//sketch 10_04_webserver_touch_auto

#include <WiFi.h>
#include <WebServer.h>
#include <ESPmDNS.h>
#include "index.h"

const char* ssid = "my network name";
const char* password = "my password";
const char* hostname = "esp32";

const int touchPin = 13;

WebServer server(80);
```

```
void handleRoot() {
  server.send(200, "text/html", index_template);
}

void handleTouch() {
  server.send(200, "text/plain", String(touchRead(touchPin)));
}

void setup(void) {
  Serial.begin(9600);
  connectWiFi();
  if (MDNS.begin(hostname)) {
    Serial.println("Webserver started");
  }
  server.on("/", handleRoot);
  server.on("/touch", handleTouch);
  server.begin();
  Serial.print("Open your browser on http://");
  Serial.println(WiFi.localIP());
  Serial.print("this may also work: http://");
  Serial.print(hostname); Serial.println(".local");
}
```

The first thing to note is that there is a new #include line for index.h. This basically allows all the contents of index.h to be included in the main sketch file, while allowing us to keep it in a separate file. In fact, if you were to cut and paste the contents of index.h into the main sketch files the sketch would work just the same without the file index.h.

The next interesting part of the sketch is that we have a new function called handleTouch. This will be called whenever a browser requests the page "touch" from the web server. This function responds to the browser with the touch reading contained in a string.

To allow the web browser to handle this new web page the following line associates requests to the "touch" page with the function handleTouch:

```
server.on("/touch", handleTouch);
...
```

It's easy to see how this example could be modified to display temperature or other more useful sensor data on a web page.

Turning the Built-in LED On and Off from a Web Page

Now you know how to display readings on a web page, what about sending commands to a web page to make the ESP32 do something, such as turn the built-in LED on and off. This can be done in a similar way to the previous example, where the main web page is loaded in to the browser once, and then the browser is responsible for sending further web request to the server to turn the LED on and off. Figure 10-6 shows the somewhat minimal user interface to do just this.

When the On button is pressed, the ESP32's built-in LED will light and when you click Off—well you guessed it!

As with the previous example, the contents of the root page are kept in index.h.

```
char *index_template = "                                        \
<script>                                                        \
    function post_switch_status(state) {                        \
        const request = new XMLHttpRequest();                   \
        request.open('POST', '/switch');                        \
        request.send(state.toString());                         \
    }                                                           \
</script>                                                       \
```

Figure 10-6 *A web switch.*

```
<h1>Switch</h1>                                                      \
<button onClick='post_switch_status(1)'>On</button>      \
<button onClick='post_switch_status(0)'>Off</button>     \
";
```

As with the previous example, the contents of index.h are split between JavaScript code to be run on the web browser and HTML user interface elements. The user interface comprises two button tags, each of which calls the JavaScript function post_switch_status with a parameter of 1 or 0 depending on whether the LED is being turned on or off. The JavaScript function itself sends a web request to the 'switch' page, with the state (0 or 1) as the data posted to web server.

Looking at the sketch itself, the main difference from the previous sketch is the handleSwitch function:

```
//sketch 10_05_webserver_switch

void handleSwitch() {
  String stateStr = server.arg(0);
  Serial.println(stateStr);
  if (stateStr == "1") {
    digitalWrite(ledPin, LOW);
  }
  else if (stateStr == "0") {
    digitalWrite(ledPin, HIGH);
  }
  server.send(200, "text/plain", "");
}
```

This function first retries that data posted to the page as server.args(0)—0 for the first (and only) data sent. It then prints this on the Serial Monitor as a useful debugging tool before setting the GPIO pin for the LED either HIGH or LOW, depending on the data sent. Note that on the Lolin32 Lite that I used, the logic of the LED is reversed, so LOW means turn the LED on.

Although this example just turns an LED on or off, that pin could control a relay to switch something much more powerful on or off.

Connecting to a Web Service

All the examples so far have involved running a local web server on the ESP32. In this example, we will look at how the ESP32 can act a bit like a browser and use data that it fetches from the Internet to do something. In this case, to use the open

Figure 10-7 *Displaying the temperature from the open weather map service.*

weather map web service to lookup the current outdoor temperature for a particular location and display the temperature and location on an OLED display like the one used in Chapter 9 (Figure 10-7).

For this project I used a Lolin32 Lite ESP32 board. This board does not breakout pin 21, one of the default I2C pins for the ESP32 with Arduino. Fortunately, it is easy to associate the I2C SDA and SCL pins with any pins on an ESP32 board. So, in this case the wiring (show in Figure 10-7) is as follows:

- GND to GND
- VCC on the OLED display to 3V on the Lolin32
- SCL on the OLED display to 16 on the Lolin32
- SDA on the OLED display to 17 on the Lolin32

The Open Weather Maps service is free if you keep your number of requests below 1 million calls a month to their API (Application Programming Interface) but to use it, you will need to register for their free tier at: https://openweathermap.org/. Registering will give you an access key to allow you to connect to their service to get their weather data. To get your key, log in, then click on the dropdown menu on your account name and select the option My API Keys (Figure 10-8).

Create a new key by entering a name (anything will do for a name) and then clicking the Generate button. This creates a long key, that you will need to copy

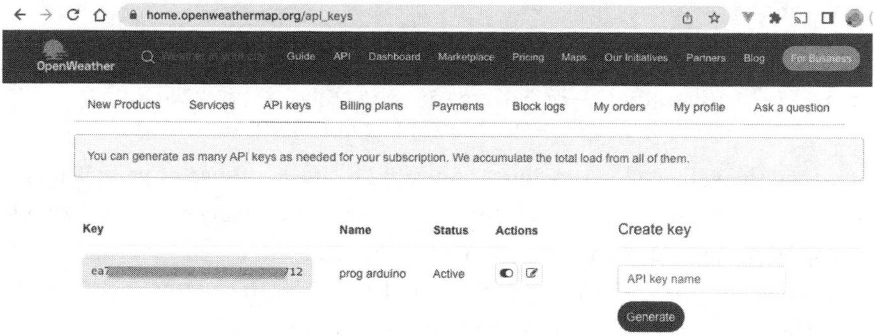

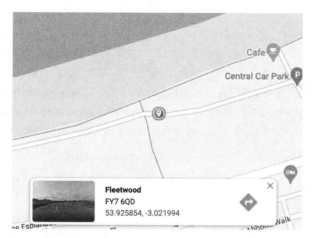

Figure 10-8 *Creating an API key.*

Figure 10-9 *Using Google Maps to find latitude and longitude.*

and paste on the sketch. You need to specify a latitude and longitude for the weather API. One way to find this information is to open Google Maps in your browser and pick a location. When you click on it, the latitude and longitude will pop-up (Figure 10-9).

The code is quite long, so open the sketch 10_06_weather_api while its operation is described; there are also the usual changes you will need to make to the sketch for your network and also the key.

```
const char* ssid = "my network name";
const char* password = "my password";
const char* url =
"http://api.openweathermap.org/data/2.5/weather?lat=
53.925854&lon=-3.021994&appid=ea751fc712f28759e8a97613b712";
```

Change the values of ssid and password to march those of your WiFi network and then paste in new values for lat and long and replace the long string for the appid at the end of the url variable with the key that you generated earlier. You can test out this URL by copying and pasting it into the address bar at the top of your browser. You will get a response something like this:

```
{"coord":{"lon":-3.022,"lat":53.9259},"weather":[{"id":802,
"main":"Clouds","description":"scattered clouds","icon":
"03d"}],"base":"stations","main":{"temp":282.52,"feels_like":
279.92,"temp_min":281.54,"temp_max":283.84,"pressure":1019,
"humidity":67,"sea_level":1019,"grnd_level":1018},"visibility":
10000,"wind":{"speed":5.08,"deg":75,"gust":6.31},"clouds":{"all":
28},"dt":1650881760,"sys":{"type":1,"id":1411,"country":"GB",
"sunrise":1650862127,"sunset":1650915062},"timezone":3600,"id":
2649312,"name":""Fleetwood","cod":200}
```

This response from the API is in a format called JSON. It is made up of attribute and value pairs grouped together in curly braces and separated by commas. We will need to extract the *temp* and *name* values from this. Continuing with the code:

```
const long fetchPeriod = 60000L; // milliseconds long lastFetchTime = 0;

Adafruit_SSD1306 display(128, 64, &Wire, -1);

int tempC = 0;
String placeText = String("Looking up..");
```

The variables fetchPeriod and lastFetchTime are used to control how often requests take place to the API and the display is updated. The global variables tempC and placeText will be used to hold the values that we want to display.

```
void setup(void) {
  Wire.begin(17, 16); // SDA, SCL
  display.begin(SSD1306_SWITCHCAPVCC, 0x3c);
  display.setTextColor(WHITE);
  display.setTextSize(1);
  Serial.begin(9600);
  connectWiFi();
  getWeatherData();
  updateDisplay();
}
```

In the `setup` function the display is initialized. Serial is also started, but is only used for test messages. As with our other network-related sketches, we call `connectWiFi`; but, if you go and look at this function, you will see that rather than writing progress messages to the Serial Monitor, these are shown on the OLED display. Finally, setup calls `getWeatherData` and then `updateDisplay`, which we will come to in a minute.

```
void loop(void) {
  long now = millis();
  if (now - lastFetchTime > fetchPeriod) {
    lastFetchTime += fetchPeriod;
    getWeatherData();
    updateDisplay();
  }
}
```

The `loop` function calls `getWeatherData` and `updateDisplay` if the `fetchPeriod` has elapsed. Now we come to the crucial call to the API contained in the `getWeatherData` function.

```
void getWeatherData() {
  if (WiFi.status() != WL_CONNECTED) {
    connectWiFi();
  }
  HTTPClient client;
  client.begin(url);
  int responseCode = client.GET();
  if (responseCode == HTTP_CODE_OK) {
    String data = client.getString();
    String tempText = extractValue(data, "\"temp\":", false);
    tempC = tempText.toInt() - 273;
    placeText = extractValue(data, "\"name\":", true);
  }
}
```

This function first checks to see if the board is connected to WiFi and if it isn't, it calls `connectWiFi`. It then creates a client connection using the url variable we defined earlier. The web request to the API is actually made when `client.getString` is called. To extract the temperature and location name from the data, the function `extractValue` is used and in the case of the temperature, the string result is converted into an `int` using the `toInt` method. The `extractValue` function is a useful function that you might want to use in your own IoT projects.

```
String extractValue(String data, char* key, boolean isString) {
   int valueStartIndex = data.indexOf(key);
   int n = strlen(key);
   String value = "";
   if (valueStartIndex > -1) {
      valueStartIndex += n;
      int valueEndIndex = data.indexOf(",", valueStartIndex);
      if (isString) {
         valueStartIndex ++;
         valueEndIndex --;
      }
      value = data.substring(valueStartIndex, valueEndIndex);
      Serial.println(value);
      return value;
   }
   return String("");
}
```

The extractValue function uses the indexOf method of String to find the start of the key within the body of the text. The length of the key is added to valueStartIndex to find the first position of the value. It then uses indexOf again to find the end of the data value (searching for a comma). The second optional parameter to indexOf is the position to start from.

One subtlety of this function is the isString parameter. If this is set to true then an extra character is removed from either end of the string by increasing valueStartIndex by 1 and decreasing valueEndIndex by 1. This removes the quotation marks from around the string.

The updateDisplay function displays the temperature and location information, by positioning the cursor and using different sized fonts.

```
void updateDisplay() {
   display.clearDisplay();
   display.setCursor(0, 0);
   display.setTextSize(6);
   display.print(tempC);
   display.setTextSize(2);
   display.print("C");
   display.setCursor(0, 54);
   display.setTextSize(1);
   display.println(placeText);
   display.display();
}
```

This project could easily be adapted to display other information about the current weather, or even display information from other APIs provided by Open Weather Maps such as weather forecast information.

Conclusion

In this chapter, we have written Arduino sketches that allow a WiFi-equipped board to both act as a web server on our local network and call web services over the Internet. Much of this code can be readily adapted to your own projects and should give you a firm basis to work from.

INDEX